RECOMBINANT
DNA TECHNOLOGY

RECOMBINANT
DNA TECHNOLOGY

Keya Chaudhuri

The Energy and Resources Institute

ISBN 978-81-7993-320-6

Suggested citation

Chaudhuri, Keya. 2013. *Recombinant DNA Technology.* New Delhi: TERI

Published by
The Energy and Resources Institute (TERI)

TERI Press	**Tel.** 2468 2100 or 4150 4900
Darbari Seth Block	**Fax** 2468 2144 or 2468 2145
IHC Complex, Lodhi Road	India +91 • Delhi (0)11
New Delhi – 110 003	**E-mail** teripress@teri.res.in
India	**Website** www.teriin.org

Printed in India

Preface

Recombinant DNA technology is a branch of molecular biology that deals with the joining of DNA molecules from two different sources and inserted into a host organism to produce a number of copies of the new genetic combinations which find applications in science, medicine, agriculture, and industry. This technology paved the way to isolate a gene, determine the nucleotide sequences, study gene transcription and protein expression, mutate genes specifically apart from many other applications. Recombinant DNA technology is not merely a collection of techniques but also an evolving science that gives us an insight into living organisms at the molecular level. The technology developed using the principles has given birth to many new streams—genomics, transcriptomics, system biology, and so on.

While teaching and taking interviews, the author came across the following statement several times: "Recombinant DNA technology is a mere technique; how would you address scientific problems with this!" This statement is the inspiration behind writing this book. The primary purpose of *Recombinant DNA Technology* is to make the current state of knowledge on the principles of recombinant DNA technology and its applications understandable to students, teachers, and scientists. With this book, the author hopes that students will be able to readily grasp the fundamental principles and themes presented in several chapters as well as practical applications of those themes, which have been illustrated in later chapters.

The author would like to thank her PhD students, especially Avirup Dutta and Sanjit Mukherjee, for their feedback, which helped her improve the book's content. She is especially grateful to Avirup Dutta, who transformed all the hand-drawn figures into print-friendly digital form. The author is also thankful to the members of her group at the Molecular and Human Genetics Division, Indian Institute of Chemical Biology, Kolkata, for providing technical and other support. Finally, she is grateful to her family members for their kind patience, support, and all possible help during the period she was busy with this project.

Author gratefully extends thanks to all original contributors and all owners of rights whose illustrations/pictures/figures/works have been adapted/reproduced in this book with citation of the source/reference. A special thanks goes to TERI Press for taking the initiative to publish the book promptly. The author is indebted to all the anonymous reviewers for their valuable inputs. Finally, the author hopes that this book may provide a new and excellent insight into the world of recombinant DNA technology.

Contents

Abbreviations

A	adenine
ADA	adenosine deaminase
AdoMet	S-adenosyl methionine
AFLP	amplified fragment length polymorphism
AMD	age-related muscular degeneration
AMV	avian myeloblastosis virus
AP	apyrimidinic
ARS	autonomously replicating sequence
ATP	adenosine triphosphate
BAC	bacterial artificial chromosome
BAP	bacterial alkaline phosphatase
BHI	biosynthetic human insulin
bp	base pairs
Bt	*Bacillus thuringiensis*
C	cytosine
C1-INH	C1 inhibitor
CCD	charge coupled device
cDNA	complementary DNA
CE	capillary electrophoresis
CIP	calf-intestinal phosphatase
CODIS	combined DNA index system
dATP	deoxyadenosine triphosphate
dCMP	deoxycytidine monophosphate
dCTP	deoxycytidine triphosphate
dGTP	deoxyguanosine triphosphate
dTTP	deoxythymidine triphosphate
ddATP	2′,3′-dideoxyadenosine-5′-triphosphate

ddCTP	2',3'-dideoxycytidine-5'-triphosphate
ddGTP	2',3'-dideoxyguanosine-5'-triphosphate
ddNTP	dideoxynucleoside triphosphate
ddTTP	2',3'-dideoxythymidine-5'-triphosphate
DDW	double distilled water
DGGE	denaturing gradient gel electrophoresis
DMS	dimethyl sulphate
DNA	deoxyribonucleic acid
dNTP	deoxynucleotide triphosphate
dsRNA	double-stranded RNA
DTT	dithiothreitol
dUTP	deoxyuridine triphosphate
EDTA	ethylene diamine tetra acetic acid
EGFR	epidermal growth factor receptor
EMBL	European Molecular Biology Laboratory
EPO	erythropoietin
G	guanine
G-CSF	granulocyte colony stimulating factor
GM-CSF	granulocyte macrophage colony stimulating factor
GMO	genetically modified organism
GST	glutathione S-transferase
HA	human influenza haemagglutinin
HANE	hereditary angioneurotic edema
HBV	hepatitis B virus
HIV	human immunodeficiency virus
IS	insertion sequence
kb	kilobase
kDa	kilo dalton
LPA	linear polyacrylamide
M	molarity
MAC	mammalian artificial chromosome
Mb	megabase
MBP	maltose binding protein
MCS	multiple cloning sites
mg	milligram

MgCl$_2$	magnesium chloride
min	minute
ml	millilitre
mM	millimolar
MMLV	moloney murine leukaemia virus
mRNA	messenger RNA
NAD	nicotinamide adenine dinucleotide
NAH	nucleic acid hybridization
NTA	nitriloacetic acid
ORF	open reading frame
PAGE	polyacrylamide gel electrophoresis
PCR	polymerase chain reaction
PFGE	pulsed field gel electrophoresis
pM	pico mole
POP6	performance optimized polymer 6
PPI	peptidyl prolyl trans isomerase
PTGS	post-transcriptional gene silencing
PTH	parathyroid hormone
Q-PCR	quantitative real-time PCR
rDNA	recombinant DNA
RdRP	RNA-dependent RNA polymerase
RF	replicative form
RFLP	restriction fragment length polymorphism
RISC	RNA-induced silencing complex
RNA	ribonucleic acid
RNAi	RNA interference
RT-PCR	reverse transcription-polymerase chain reaction
SCID	severe combined immunodeficiency
SDF	stromal cell-derived factor
SDM	site-directed mutagenesis
SDS-PAGE	sodium dodecyl sulphate-polyacrylamide gel electrophoresis
siRNA	small interfering RNA
SNP	single nucleotide polymorphism
SSCP	single-strand conformation polymorphism
STM	signature-tagged mutagenesis

STR	short tandem repeats
SUMO	small ubiquitin-like modifier
T	thymine
TAE	Tris-acetate-EDTA
Taq	*Thermus aquaticus*
TdT	terminal deoxynucleotidyl transferase
TE	Tris-EDTA
TPA	tissue plasminogen activator
tRNA	transfer RNA
Tris	Tris(hydroxymethyl)aminomethane
Trx	thioredoxin
U	uracil
UV	ultraviolet
V	volts
VEGF	vascular endothelial growth factor
VNTR	variable number of tandem repeats
YAC	yeast artificial chromosome
YEP	yeast episomal plasmid
YIP	yeast integrative plasmids
YRP	yeast replicative plasmids
α	alpha
β	beta
μl	microlitre
μg	microgram

1

Recombinant DNA Technology

OBJECTIVES

After reading this chapter, the student will be able to:

• Understand the importance of recombinant DNA

• Explain the basic procedure of recombinant DNA or rDNA technology

• Describe the impact of recombinant DNA

INTRODUCTION

The deoxyribonucleic acid (DNA) molecule of a living organism keeps all the information needed for its function. Chemically, the DNA molecule is known as the "double helix", consisting of backbones made with sugars and phosphates joined by ester bonds. A nitrogenous base is attached to each sugar. There are only four types of bases: adenine (A), cytosine (C), guanine (G), and thymine (T). All organisms contain the same four bases, but the sequence of bases along the backbone of DNA molecule and the number of bases present therein vary and create diversity. The information coded in DNA is first transcribed into mRNA, which is further translated into protein, and proteins are the key molecules carrying out all functions. Thus a different protein can be made by changing the DNA sequence. Scientists have made use of this fact to create a new technology called recombinant DNA, or rDNA, technology.

rDNA is the general name for artificially combining DNAs from two or more different sources to create a new DNA molecule. It is sometimes known as "chimera". The most common rDNA technology involves the introduction of a fragment of DNA molecule from one organism to a second organism. The rDNA molecule, when introduced into a different organism, can code for a specific trait, which was not found earlier. It differs from genetic recombination in that it does not occur naturally through the processes inside the cell but is engineered.

An rDNA molecule can be prepared in vitro in a laboratory test tube and introduced into an organism. A vehicle called a vector is required for this. In this manner, the genetic make-up of an organism can be altered. Using rDNA technology, many human proteins have been synthesized in bacteria, yeast or cultured mammalian cells. A human gene cloned in bacteria or yeast can produce large quantities of the protein of choice. More than hundreds of proteins are manufactured with the help of rDNA technology and have useful applications in human diseases. Some examples are listed in Table 1. A recombinant protein can be created using rDNA technology. A useful example is the synthesis of human insulin in bacteria using rDNA technology.

The rDNA experiment was first conducted by Boyer and Cohen in 1973. They devised techniques to isolate plasmids (vehicles for rDNA construction and transfer to a suitable host), cut them at precise locations, recombine them with a foreign DNA by ligase, and finally insert them into another cell with precision thereby creating a recombinant or transgenic bacterium. In this way, bacterial cells were used to produce the desired protein. The technology was a major breakthrough for genetic engineering. However, rDNA technology was made possible by the discovery of restriction endonucleases by Werner Arber, Daniel Nathans, and Hamilton Smith, for which they received the Nobel Prize in Medicine in 1978.

Table 1 Some human proteins synthesized by rDNA technology used in the treatment of human diseases	
Recombinant human protein	*Diseases treated*
Insulin	Diabetes mellitus Type I
Factor VIII	Haemophilia A
Factor IX	Haemophilia B
Somatotropin	Growth disorder
Tissue plasminogen activator (tPA)	Heart attack
Adenosine deaminase (ADA)	Severe combined immunodeficiency (SCID)
Parathyroid hormone (PTH)	Hypoparathyroidism
Interleukins	Immune disorders, cancers
C1 inhibitor (C1INH)	Hereditary angioneurotic edema (HANE)
Erythropoietin (EPO)	Anaemia, kidney disorder
Epidermal growth factor	Ulcers, burns
Interferons (•,•,•)	Various cancers, immune disorder, viral infection
Superoxide dismutase	Free radical damage in kidney transplant
A1-Antitrypsin	Emphysema
Granulocyte macrophage colony stimulating	Stimulation of bone marrow after bone marrow transplant factor (GM-CSF)
Granulocyte colony stimulating factor (G-CSF)	Stimulating neutrophil production after chemotherapy and for mobilizing hematopoietic stem cells from the bone marrow into the blood

IMPORTANCE OF RECOMBINANT DNA

rDNA technology, the science behind transgenic animals, insect-resistant crops, and artificial and other genetically modified products, has gained considerable importance over the past decade, particularly to meet the growing demands of the twenty-first century. With a constant decrease in agricultural areas and increase in genetic diseases, rDNA technology is increasingly becoming the source of solution for these. Some of the areas where rDNA technology is having a considerable impact are mentioned as follows.

Genetically modified organisms (GMOs)

❖ Transgenic animals as experimental models in biomedical research.

❖ Transgenic fruit flies (*Drosophila melanogaster*) as model organisms used in biomedical research to study the effects of genetic changes on development.

Agriculture

❖ Better crops (resistant to insects, pests, herbicides, and harsh environmental conditions such as drought, high salt, and heat or cold).

❖ Plants that produce their own insecticides.

❖ Crops with improved product shelf life.

❖ Crops with increased nutritional value.

❖ Crops resistant to virus damage.

Medicine

❖ Recombinant vaccines (hepatitis B).

❖ Prevention and cure of sickle-cell anaemia.

❖ Prevention and cure of cystic fibrosis.

❖ Production of clotting factors.

❖ Production of insulin.

❖ Production of recombinant pharmaceuticals.

❖ Germ line and somatic gene therapy.

BASIC PROCEDURE OF RECOMBINANT DNA TECHNOLOGY

The organism used to donate DNA for analysis is called a donor organism. A carrier molecule called a vector, which is an autonomously replicating molecule such as a circular plasmid DNA, is needed. The basic procedure of rDNA technology (Plate 1) is briefly described as follows.

❖ *Isolation of DNA* DNA is isolated from the vector and the source containing the gene of interest. Bacterial plasmids that are autonomously replicating circular DNA molecules may serve as vectors. They also contain markers (for example, *amp*^{r} and *lacZ*) for recognition and further selection.

❖ *Restriction digestion*   The vector DNA is cut with a specific restriction enzyme at the precise location to open it up. The foreign DNA is cut into a number of fragments, one of which contains the gene of interest. When the restriction enzyme cuts, it produces sticky ends on the plasmid as well as on the foreign DNA (Figure 1).

❖ *Ligation*   The cut plasmid DNA molecules are mixed with foreign DNA fragments, and a ligation reaction is set up. The sticky ends of the foreign DNA fragment base pair with the complementary sticky ends of the plasmid DNA. The enzyme DNA ligase catalyses the formation of covalent bonds and joins the foreign DNA and the plasmid vector. Self-ligation is also possible at this stage but can be carefully controlled. The vector containing the foreign DNA fragment is called rDNA because it consists of a novel combination: the DNA from the donor organism (which may be any species) with the DNA from a vector (a plasmid or a phage).

❖ *Introduction of rDNA into bacterial cells*   The naked rDNA is used to transform bacterial cells. It is common for a single rDNA molecule to enter a bacterial cell. The recombinant plasmid will replicate within the bacterial cell by using the enzymatic machinery of the host. A transformed cell with the rDNA will then divide and multiply and form a colony with millions of cells, each of which carries the rDNA molecule.

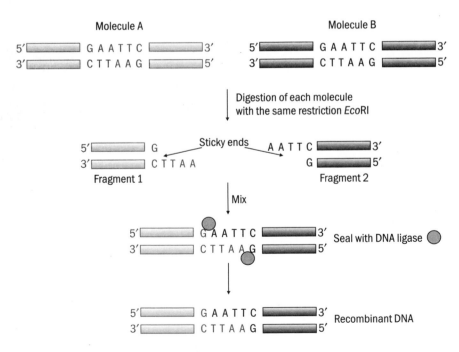

Figure 1   Basic principle of restriction digestion

Bacterial colonies containing the recombinant plasmid are selected on the basis of markers present in the vector. An individual colony will contain a large population of identical insert DNA and is called a *DNA clone*. Much of the cloned DNA fragment analysis can be performed when it is in the bacterial host. Later, however, it is often desirable to reintroduce the cloned DNA into the cells of the original donor organism to carry out specific manipulations of genome structure and function.

❖ *Screening of recombinants* Once a recombinant plasmid is transformed, the bacterial cell that has picked up the recombinant clone grows into a colony. Cells containing various plasmids will grow into separate colonies. One has to search through these colonies to find out which one contains the desired foreign DNA sequences. This can be done by use of replica plating and hybridization as illustrated in Figure 2.

At first, a suitable number (usually of low density for convenience of screening) of colonies are grown on the nutrient plate. Replica plating is then used to transfer these colonies to a number of plates. This technique prepares a number of culture plates containing representative colonies at the same position on each plate. One of the replica plates is then used to further screen for the desired DNA. Bacterial cells are transferred

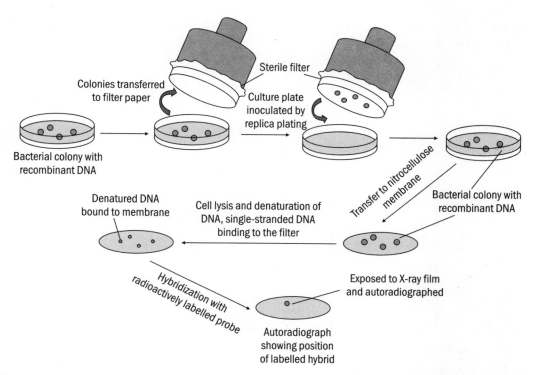

Figure 2 Basic procedure for identification of a bacterial colony containing desired foreign DNA fragment

to a solid matrix, usually a nitrocellulose membrane, and after lysis, the DNA is fixed onto the membrane. The DNA is further denatured and the membrane is incubated with radioactively labelled single-stranded DNA probe. The probe DNA contains sequences complementary to the desired foreign DNA fragment being sought. The membrane is then washed to remove the non-hybridized probes. In order to determine the location of the labelled hybrids, the membrane is placed on an X-ray film and autoradiographed. The colonies that hybridize with the probe will produce black spots on the X-ray film after development. The X-ray film is then compared with the master plate to find out which colonies contain the specific sequence complementary to the probe.

The complete process of rDNA technology can be schematically depicted (Figure 3).

The term rDNA should be distinguished from natural DNA recombinants. Natural DNA recombinants result from the crossing over between homologous chromosomes in eukaryotes as well as prokaryotes and occur in vivo. rDNA, on the other hand, is an unnatural union of DNAs from non-homologous sources and can be from any organism and is done in vitro. These rDNAs are also termed chimeric DNA, after the mythological Greek monster Chimera. The chimera has been symbolized as an impossible biological union, a combination of parts of different animals. Likewise, rDNA is a DNA chimera and would be impossible without experimental manipulation, called rDNA technology.

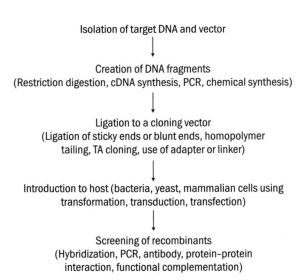

Figure 3 Overview of basic procedure of recombinant DNA technology

IMPACT OF RECOMBINANT DNA

The rDNA technology is starting to have a major impact on our day-to-day life. But like all major advancements in science, there is a flip side to it as well. This great technology, developed for the benefit of mankind, has the potential to be one of its worst enemies depending on its use. There is a grey area that is still not clearly defined, as each individual has his/her own way of defining good and bad, right or wrong. Based on the standard beliefs of our society, the pros and cons of this powerful technology, as it can influence our lives, are listed as follows. It is definitely a matter of opinion of what is good or bad, but the impact it will have on one's life will help one have a better understanding of rDNA technology.

Box 1   Recombinant DNA technology in the synthesis of human insulin

One of the biggest breakthroughs in recombinant DNA technology occurred in the manufacture of biosynthetic "human" insulin. The insulin hormone was discovered in 1921. Thereafter, diabetic patients with elevated sugar levels due to impaired insulin synthesis could be treated with insulin, which was usually isolated from the pancreatic glands of bovine or porcine animals. Bovine or porcine insulin is similar in composition to human insulin, but has some variations. So the immune systems of some patients produced antibodies against it, neutralizing its actions and resulting in inflammatory responses at the injection sites. Long-term injection of foreign insulin also gave rise to many complications. Therefore, scientists began to search for alternative sources.

The first genetically engineered human insulin was synthesized in the laboratory of Herbert Boyer in Genetech, USA in 1977. The human insulin gene was inserted in a suitable plasmid vector and introduced into *Escherichia coli* successfully. The genetic regions coding for A and B chains of insulin were cloned separately. The *E. coli* cells produced A and B chains that were later joined by chemical techniques (Plate 2). The final product was chemically identical to its naturally produced counterpart in humans. The bovine insulin, which was used earlier, differed from human insulin in a few amino acids. When injected, it was recognized as a foreign antigen, and antibodies were produced against it, and its effect was nullified after a few days. Creation of human insulin could avoid this problem.

A major difficulty of creating insulin with *E. coli* was that chemical methods were required in addition to the genetic methods for rejoining A and B chains. Later on, gene coding for human proinsulin was inserted into *E. coli* cells, which was then grown by fermentation to produce proinsulin. This was then processed enzymatically to cleave the connecting peptide and convert proinsulin to insulin. Nowadays, human insulin is synthesized in yeast cells, which secrete insulin. The yeast-derived insulin is completely identical to human insulin.

Pros

❖ Improved, safer, and cheaper medicines (such as insulin).

❖ Improved disease-resistant livestock.

❖ Better crops (with improved shelf life, nutritional value, and resistance to diseases).

❖ Prevention of genetic diseases.

❖ Treatment for pre-existing conditions (such as cancer).

Cons

❖ Development of multi-drug-resistant bacteria and viruses developing antibiotic resistance, developing resistance to fungi may lead to safety and environmental concerns.

❖ Ethical dilemmas over human treatment.

❖ Doctors may start using patients as experimental models.

❖ As man tries to play God, germ line treatment, instead of treating diseases, may turn into a method for creating customized babies raising serious ethical issues.

❖ Genetically modified crops may cause health hazards, for example, Bt brinjal (Box 2).

Box 2   BT brinjal controversy

Bt brinjal is a genetically modified or a transgenic brinjal developed by Monsanto, an American company, and it is being marketed in India by Mahyco. This variety of brinjal plant has been genetically engineered to become resistant to lepidopteron insects, such as brinjal fruit and shoot borer (*Leucinodes orbonalis*) and fruit borer (*Helicoverpa armigera*). The gene coding for Cry toxin of the soil bacterium *Bacillus thuringiensis* is added to brinjal, and hence the name Bt brinjal. The insecticidal properties of Bt arise mainly due to the action of this Cry toxin. The primary action of this Cry toxin is to lyse the midgut epithelial cells of the target insects. The *Cry1Ac* gene is inserted into the brinjal cells in young cotyledons through an Agrobacterium-mediated vector. When the Cry protein is expressed in brinjal plants, the latter becomes resistant to insect attacks.

Although the concept of Bt brinjal sounds appealing, several studies have shown that there are potential health hazards with the bioengineered crops. Tests show that Bt brinjal has 15% less calories and contains an insecticide toxin with different alkaloid content as compared to its natural counterpart. Animal experiments indicate significant differences in blood chemistry based on sex of the animal or period of

---

Contd...

Box 2 Contd...

measurement. Effects were also observed on blood clotting time (prothrombin), total bilirubin (liver health), and alkaline phosphate counts in goats and rabbits, posing as a major environmental and biodiversity threat. Increased water consumption with decrease in liver weight and liver-to-body weight was also observed in Bt brinjal-fed rats.

Not only Bt brinjal but other genetically modified crops like Bt cotton have shown ill effects on human health with reports of allergies, itching skin, eruptions on the body, swollen faces, on exposure to Bt cotton. Organizations like Greenpeace have been strongly opposing the genetically modified variety of brinjal. All these reports and controversies have led the Government of India to postpone the release of Bt brinjal in India and declare that it needed some more time before it could be commercially available.

SUMMARY

- rDNA is the general name for artificially combining DNAs from two or more different sources to create a new DNA molecule. Many human proteins have been synthesized with the help of rDNA technology and have useful applications in human diseases. rDNA technology has been utilized in creating genetically modified organisms and in creating better crops.

- rDNA molecules are generated in vitro by the insertion of DNA fragments from any organism into vector DNA molecules. The DNA fragments are generated by the digestion of DNA with restriction enzymes that cleave it at specific locations. Joining of foreign DNA fragments to vector molecules is done by an enzyme ligase. The rDNA molecules are then introduced into host cells by transformation where they replicate and produce rDNA molecules in large numbers.

- The presence of foreign DNA in a recombinant clone can be screened by hybridization using polynucleotide probes.

REVISION QUESTIONS

1. Is it theoretically possible for a gene from any organism to function in any other organism? How?

2. List the basic steps involved in genetic engineering and describe what each component does. Draw a picture showing the various steps involved in cloning the insulin gene.

3. Name five things you can do with cloned DNA and explain the importance of each.

4. What is a cloning vector? What are its essential characteristics?

5. Why do you need to attach DNAs to vectors to clone them?

2

Methods for Creating Recombinant DNA Molecules

OBJECTIVES

After reading this chapter, the student will be able to:
- Explain sources of DNA for cloning
- Discuss the selection of host and vector
- Describe the commonly used vectors
- Analyse plasmids as vectors
- Describe the preparation of vector and insert DNAs
- Explain restriction enzymes
- Describe joining of DNA molecules
- Understand propagation of plasmids carrying foreign molecules

INTRODUCTION

Studies of the structure and function of a gene at the molecular level require large quantities of individual genes in pure form. This can be achieved by recombinant DNA, or rDNA, technology, wherein the gene of interest is first inserted into a vector and then transformed to a select host bacterium and replicated therein. This process is termed cloning of a gene. A recombinant DNA is simply any DNA molecule composed of sequences derived from different sources. The basic scheme for creating rDNA molecules was presented in the previous chapter. In this chapter, each step will be discussed in detail.

SOURCES OF DNA FOR CLONING

Sources of DNA for cloning, normally known as inserts, include the following.
❖ Genomic DNA (full gene, part of a gene, or a large genomic fragment).

❖ Complementary DNA (cDNA) of an mRNA.
❖ Polymerase chain reaction (PCR) product, that is, a fragment of DNA amplified through PCR.
❖ Chemically synthesized oligonucleotides.

SELECTION OF HOST AND VECTOR

The most favoured host is the bacterium *Escherichia coli*, which has several advantages that make it amenable for rDNA technology. The cells grow fast in a simple, inexpensive growth medium and the generation time (or doubling time) is just 20–30 min. Its genetics is well understood, the gene transfer mechanisms are easy to perform, and the strains used for cloning are genetically engineered so that they are relatively harmless to humans. It can, therefore, be easily manipulated. Extrachromosomal copies of DNA (plasmids and bacteriophage DNA) can be exploited to carry foreign DNA fragments.

The disadvantages of using *E. coli* are as follows.

❖ Post-translational modification of many proteins is not possible (for example, glycosylation, acetylation).
❖ Its tendency to aggregate any protein expressed at high level and form insoluble products.
❖ Lack of secretion of foreign proteins.
❖ Lack of splicing, that is, processing of non-coding or intron sequences.

Some of these problems can be overcome by using eukaryotic hosts, such as yeast or mammalian cells. The advantages and disadvantages of these systems are summarized in Table 1.

Table 1 Advantages and disadvantages of using eukaryotic hosts

Host System	Advantages	Disadvantages
Yeast	Processing of RNA transcripts is possible so that introns can be removed efficiently	Often make differences in glycosylation of proteins compared to the original eukaryotic source
	Glycosylation of proteins is possible	Refolding often needed
	Can be grown in low-cost media and generation time is low	Protein production lower than that obtained from bacteria
	Secretes protein products Non-pathogenic	
Mammalian cells	Ability to process RNA transcripts properly for gene expression	
	Ability to modify the expressed proteins by cleavage, glycosylation, refolding Secretion is also possible	

A vector is an autonomously replicating DNA molecule that can be used to carry foreign DNA fragments. All vectors are based on naturally occurring DNA sequences that can be replicated under particular circumstances. Most commonly used vectors are based either on plasmids or bacteriophage lambda (λ). A vector used predominantly for reproducing DNA fragments is often referred to as a cloning vector. If it is used for expressing a gene contained within the cloned DNA, it is called an expression vector. A vector must possess the ability to self-replicate and must contain selectable markers so that recombinant cells can be distinguished from non-recombinant cells. Some commonly used vectors are described as follows.

Bacteriophage Vectors

Bacteriophages are viruses that infect bacteria and are used as cloning and expression vectors. Examples are M13, lambda, and P1 phages of *E. coli*.

M13 is a single-stranded DNA containing filamentous phage, which infects only male bacterial cells after attachment to the F pilus. Inside the host, M13 forms a double-stranded DNA during replication (called replicative form, RF). This RF DNA can be isolated from the cells like plasmids, and a foreign DNA is inserted into it. This chimeric DNA is then returned to the host cell as plasmids.

Bacteriophage lambda has been studied extensively and various cloning and expression vectors have been developed using lambda phage. Lambda phage infection can result in lysis of the host or in lysogeny. Lysis means production of a large number of phage particles, which burst open the host leading to its death. Lysogeny, on the other hand, signifies integration of viral DNA into the host genome, leading to the persistence of viral prophage in the host with minimal viral gene expression. Lambda phage contains a linear DNA of 45 kb with a 5′ extension of 12 bases (termed *cos* site) at both ends. These extensions, being complementary to each other, anneal to form circular genome. About 23 kb region of lambda DNA does not contain essential genes. This region can, therefore, be replaced by foreign DNA to be cloned.

Another *E. coli* phage P1 has been utilized for vector construction. P1 phages, like lambda, have both lytic and lysogenic cycles in their life. However, during lysogeny, unlike other lysogenic phages, P1 can exist in the bacterial cells independently, just like plasmids. During cell division, P1 existing as plasmid is equally partitioned into two daughter cells. P1 vectors can hold about 110 kb inserts.

Yeast Vectors

There are several types of yeast vectors, such as yeast episomal plasmids (2 μm), yeast integrative plasmids (YIP; yeast chromosome having bacterial plasmid integrated into it), yeast replicative plasmids (YRP; any circular duplex DNA containing 100-base-pair autonomously replicating sequence, or ARS, of yeast and can be used for cloning in yeast), and yeast artificial chromosome (YAC).

Yeast episomal plasmid (YEP) vector contains the origin of naturally occurring 2 μm plasmid of yeast and three REP genes—*REP1*, *REP2*, and *REP3*. *REP1* and *REP2* genes (*trans*-acting) promote partitioning of the plasmid during cell division, and *REP3* regulates *REP1* and *REP2*. The copy number of YEP is about 50–100 per cell.

YAC can accept large fragments of foreign DNA (usually 1 Mb). These vectors contain ARS of yeast and centromeric and telomeric sequences so that the rDNA can be maintained as a yeast chromosome and useful selectable marker.

The size of the insert DNA that can be accommodated in different vectors is summarized in Table 2.

Shuttle Vectors

Shuttle vectors are specifically designed to replicate in two different hosts, such as bacteria and yeast or two different bacterial species or bacteria and mammalian cell, so that DNA can be transferred between these two. The advantage of using shuttle vectors is that DNA manipulation can be done easily in one host and then shifted to the more difficult to manipulate the host. Yeast shuttle vector is most commonly used. The shuttle vector is constructed in such a way so as to accommodate the origin of replication of each host. In yeast shuttle vector, the origin of replication and antibiotic resistance marker (bla for ampicilin resistance) is usually kept as *E. coli* component. An ARS, a CEN, a yeast centromere, and a selectable marker for yeast (URA3 or gene coding for uracil synthesis) are included as yeast components.

Methods used to create rDNA molecules using plasmids (Box 1) are discussed in the subsequent section.

Table 2 Size of insert DNA for different vectors	
Vector	*Maximum size insert*
Plasmid	<10 kb
Phage λ	–5–20 kb
Cosmid	35–45 kb
Bacterial artificial chromosome (BAC)	75–300 kb
Yeast artificial chromosome (YAC)	100–1000 kb (1 Mb)
Mammalian artificial chromosome (MAC)	100 kb to > 1 Mb

Source Allison (2006)

Box 1 Plasmids

Plasmids occur naturally in bacteria as extrachromosomal DNA molecules. Plasmids can replicate independent of chromosomal DNA and contribute to the function of the cell. The existence of plasmids has now been reported in all kingdoms of life—archaea, bacteria, and eukaryotes like yeast. In nature, plasmids vary in size, structure, copy number, mode of replication, ability to propagate in different bacteria, and in the phenotypic traits they carry.

The size of a plasmid varies from 1 kilobase pair (kbp) to about 1000 kbp. The copy number can vary from 1 to about 100 or more. Plasmids that exist in <20 per bacterial cell under normal growth conditions are regarded as high copy number plasmids.

Types

Plasmids have been grouped in various ways based on their properties (Table A).

Table A Types of plasmids

Characteristics	Type of plasmid recognized
Mode of transfer from one cell to another	Conjugative, non-conjugative or mobilizable
Copy number	High copy number, low copy number
Control of replication	Relaxed, stringent
Host range	Narrow, broad
Function	R-plasmid. col, degradative
Incompatibility	Several types (~ 30 in *E. coli*)

Naturally occurring plasmids are transmitted to a new host by conjugation. In the laboratory, plasmids are usually transferred by transformation process. In general, relaxed plasmids are maintained in multiple copies under normal growth conditions, while stringent plasmids are present as a limited number of copies per cell. The stringent control of replication means that the plasmid replication is coupled to the host chromosomal replication so that only one or a few copies of the plasmid are generated in each cell, resulting in low copy number. On the other hand, the copy number of relaxed plasmids can be increased artificially if host protein synthesis is stopped by the use of chloramphenicol.

The most notable plasmid-borne function is the resistance to antibiotics carried by R-plasmids. Many bacteria become antibiotic resistant as they acquire R-plasmids. Col plasmids, such as ColE1 of *E. coli*, contain a gene coding for colicin-like proteins that are capable of killing other *E. coli* cells. The plasmid carries a second gene that confers immunity to the action of colicin so that the bacteria containing Col plasmid are protected from the lethal effect of their own product.

Contd...

Box 1 Contd...

> Incompatibility is another property unique to plasmids. When two plasmids cannot stably coexist in the same host cell in the absence of any selection pressure, they are said to be incompatible and belong to the same incompatibility group.
>
> **Nomenclature of plasmids**
>
> Plasmids are named as p(XY)(numbers or ID), where p stands for plasmid, XY are the initials of the discoverer or the representation of the discovering laboratory, and numericals are the IDs representing the isolate number. For example, pBR322 is a plasmid that was first isolated by Bolivar and Rodriguez and 322 is the lab isolate number. Another plasmid pUC denotes the plasmid designed from the University of California.

Plasmids as Vectors

Naturally occurring plasmids, especially *E. coli* plasmids, have long been employed as vehicles for propagation, manipulation, and transfer of specific genes. All plasmid vectors have the following common features: a replicator, a selectable marker, and a cloning site (Figure 1).

A replicator is required for plasmid replication. It consists of the origin of replication (*ori*) and genes encoding those replication proteins and RNAs that are plasmids encoded and required for replication. The *ori* sequences provide the replication initiation site and the anchor for the mitotic segregation of the plasmid. In *E. coli* cloning and shuttle vectors, the *ori* sequences are derived from naturally occurring plasmids such as pMB1/ColE1 in pBR322, the pET series of plasmids, or p15A in the pACYC vector series.

The plasmid must contain a selectable marker because using this marker, one can select the plasmid containing the host cell from those cells that do not contain the plasmid. For example, if a plasmid contains *bla* gene coding for an enzyme that destroys ampicillin, the host cells containing the plasmid will survive when grown in the presence of ampicillin. Cells that do not harbour the plasmid-carrying *amp*^{r} marker will be killed. This serves as a suitable basis for detection of the plasmid in a host cell. The gene encoding resistances to some antibiotics are often termed dominant markers. Other examples include the neomycin phosphotransferase gene for kanamycin resistance, the tetracycline efflux gene for tetracycline resistance, and the chloramphenicol acetyl transferase (CAT) gene for chloramphenicol resistance. Recessive markers are also used. These markers complement for an auxotrophic deficiency in the cell.

The selection of a plasmid by an antibiotic resistance alone does not tell us whether the plasmid contains foreign DNA. A second selectable marker is required for this purpose. Let us suppose a plasmid contains two antibiotic resistance markers *amp*^{r} and *tet*^{r}. If a

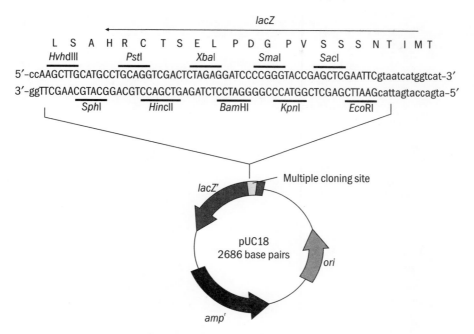

Figure 1 Typical plasmid vector

foreign DNA is inserted in such a way that the tetracycline efflux gene is disrupted, but not the *bla* or *amp*^r gene, the bacterial host containing rDNA will be sensitive to tetracycline but resistant to ampicillin. Therefore, cells containing recombinant plasmids will grow in the presence of ampicillin and will be killed when grown in the presence of both ampicillin and tetracycline. The host cells harbouring only the vector will grow significantly in the presence of both ampicillin and tetracycline. Some antibiotic resistance markers and their mode of resistance are summarized in Table 3.

The cloning site is a restriction endonuclease cleavage site into which foreign DNA can be inserted without interfering with the plasmid's ability to replicate or to confer the selectable phenotype on its host. Over the years, plasmid vectors have increased in sophistication; in addition to these three basic elements, many vectors now include features that make them particularly suitable for specific types of experiments.

Plasmids used as cloning vectors, in general, can accommodate foreign DNA fragments of about 2–10 kbp. Since the plasmids used as vectors often contain a single origin of replication, the time required for complete replication depends on plasmid size. Under selective pressure in a growing bacterial culture, plasmids containing large inserts are prone to deletion. Thus it is necessary to stick to the reasonable upper limit when cloning in a plasmid vector takes place. Many other useful vectors are available for cloning foreign DNA.

Table 3 Antibiotic resistance markers

Antibiotic	Mode of action	Gene	Resistance conferred by protein	Mechanism
Ampicillin	Inhibits the bacterial transpeptidase involved in peptidoglycan biosynthesis and thus inhibits cell wall biosynthesis	*bla*	Beta-lactamase	Cleaves the beta-lactam ring of ampicillin
Kanamycin	Interacts with three ribosomal proteins and with rRNA in the 30S ribosomal subunit, to prevent the transition of an initiating complex to a chain-elongating complex, and thus inhibits protein synthesis	*aph*	Aminophos photransferases Aph (3')-I or Aph (3')-II	Transfer phosphate from ATP to the kanamycin to inactivate it
Chloramphenicol	Inhibits the activity of ribosomal peptidyltrans-ferase, and thus inhibits protein synthesis	*cat*	Chloramphenicol acetyl transferase	Transfers an acetyl group from acetyl CoA to chloramphenicol and inactivates it
Tetracycline	Binds to a single site on the 30S ribosomal subunit to block the attachment of aminoacyl tRNA to the acceptor site and thus inhibits protein synthesis	*tetA*	TetA efflux protein Gradient	Catalyses the energy-dependent export of tetracycline from the cell against a concentration

PREPARATION OF VECTOR AND INSERT DNA

Two major types of enzymes are involved in the preparation of vector, insert, and rDNA. These are restriction endonucleases and DNA ligases. Restriction endonucleases are able to cleave the double-stranded DNA by recognizing a specific short DNA-sequence in it. DNA ligase, on the other hand, joins two pieces of DNA by the formation of phosphodiester bonds.

Restriction Enzymes

The ability to cut DNA at specific sites became possible due to the Nobel Prize winning discovery of restriction enzymes by Arber, Nathans, and Smith in 1978. The restriction enzymes (or restriction endonucleases) cleave double- or single-stranded DNA at specific recognition sequences called restriction sites. The term "endonuclease" applies to sequence-specific nucleases that break nucleic acid chains somewhere within the DNA rather than at the ends of the molecule.

The names of restriction enzymes reflect their origin. The first letter of the name comes from the genus, and the second and third letters come from the species of bacteria they were isolated from. The number following the nuclease name indicates the order in which the enzyme was isolated from the bacterial strain and is written as a Roman numeral. For example, the first restriction enzyme to be isolated from the bacterium *Providencia stuartii* was named *PstI*, and the second to be isolated from *Bacillus stearothermophilus* strain ET was named *Bst*EII.

Bacteria produce restriction enzymes as a defence against bacterial viruses or bacteriophages. Inside a bacterial host, foreign DNA is restricted, which is why the phage is unable to infect the bacteria.

The restriction enzyme cuts the DNA at the internal phosphodiester bond; for a double helix, two incisions are made, one at each strand. The bacteria avoid cutting their own genomic DNA by modifying it, particularly by methylation; the modification enzyme methylase recognizes the same target site as that of the restriction enzyme and adds a methyl group to the specific nucleotide at the restriction site, thus preventing it from restriction digestion. The bacterial DNA is methylated immediately after replication; it is unlikely that the infecting phage DNA will be methylated instantaneously. Thus the phage DNA will be an appropriate target for restriction cleavage. Modification systems vary among bacterial strains.

Restriction enzymes cleave DNA at a specific sequence called the *recognition sequence.* For example, the restriction enzyme *Eco*RI, isolated and characterized from *E. coli* strain RY13, recognizes a six-base-pair double-stranded DNA sequence 5'-GAATTC-3' and cleaves DNA on both strands between the G-base and the closest A-base residues. Once the cuts have been made, the resulting fragments are held together only by the relatively weak hydrogen bonds that hold the four complementary bases to each other. The weakness of these bonds allows the DNA fragments to separate from each other. Each resulting fragment has a protruding 5'-end composed of unpaired bases. The protruding ends are called "sticky" or "cohesive" ends because they will bond with complementary sequences of bases.

The number of restriction enzymes isolated so far is more than 1000. There are three types of restriction endonucleases of which type II restriction enzymes are widely used in the construction of rDNA molecules.

In order to join the vector molecule with a foreign DNA fragment, both of these should be cleaved in such a way that complementary ends are generated. For example, if the *Eco*RI enzyme is used to digest both the vector and insert DNA, complementary ends are generated, and the vector molecule will find its complementary end at the insert, and vice versa (Figure 2). Although there is a possibility that the vector molecules will rejoin, this can be controlled by the addition of excess insert. However, to join two fragments together, it is not essential that the vector and the insert be digested with the

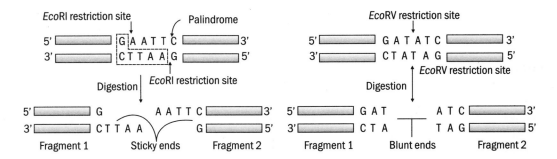

Figure 2 Cleavage by restriction enzymes

same restriction enzyme. Many different restriction endonucleases exist that produce compatible ends. Moreover, various restriction enzymes such as *Sma*I and *Alu*I have blunt ends and do not produce overhang after cleavage.

The generation of compatible, sticky-ended DNA fragments makes them associate with each other through complementary base pairing, but they will not form a continuous sugar–phosphate DNA backbone. The joining of blunt ends is done with different strategies. The strategies employed for joining vector and insert molecules are now discussed.

Processing Restriction Fragments

If a restriction digestion produces more than one fragment, it is often required to purify the fragments. The digestion product is run on a low-melting agarose gel. Low-melting agarose is a preparation of agarose having lower melting point compared to the normal agarose. The DNA fragments can be easily recovered from the gel. After running the gel, the DNA fragments are visualized with ethidium bromide under UV. The desired bands are excised using a scalpel. The DNA can be recovered from the gel by standard purification procedure.

On the other hand, if the digestion product is only one fragment or if a complex mixture is to be used for library production, it is enough to inactivate the restriction enzyme before proceeding to the next step. Normally, it is heat inactivated, or if the enzyme is heat stable, phenol extraction is required.

DNA fragments for ligation can be generated in several ways. A major procedure is by restriction digestion. Restriction digestion is specific, but random fragments can also be generated by digestion with non-specific nucleases. Apart from this, fragments can be generated by mere shearing of DNA by mechanical procedure, creating a collection of random fragments. For functional characterization of eukaryotes, it is often necessary to clone cDNAs prepared by the reverse transcription of mRNA. Polymerase chain reaction can be used to amplify a DNA fragment easily or a gene may be chemically synthesized for cloning.

JOINING OF DNA MOLECULES

Having described the methods available for cleavage of DNA molecules at some specific places, the next task of a genetic engineer is to consider the ways in which DNA fragments can be joined to artificially create rDNA molecules. In the 1960s, several laboratories simultaneously discovered DNA ligase, an enzyme that catalyses the formation of a phosphodiester bond between two DNA molecules. DNA ligase enzymes require a free hydroxyl group at the 3′-end of one DNA chain and a phosphate group at the 5′-end of the other. The formation of a phosphodiester bond between these groups requires energy. In *E. coli* and other bacteria, NAD^+ plays this role, whereas in animal cells and bacteriophages, ATP drives the reaction. DNA ligases are only able to join DNA molecules that are part of a double helix; they are unable to join two molecules of a single-stranded DNA (Plate 1).

Optimization of Ligation Reaction

In gene cloning, the ligation step is critical as there may be failure of ligation and there may be cases where self-ligation of vectors is more prominent than the ligation of foreign DNA with the vector. The presence of inhibitory agents may contaminate DNA preparation, degrade enzyme activity, or degrade DNA, leading to the failure of ligation. The other conditions to be controlled for ligation are temperature and the concentration of vector and insert DNAs. Earlier studies had used a reaction temperature of 10°C (often 4°C) and long time of incubation (usually overnight). Many protocols recommend the reaction at 16°C or room temperature. On advancement of technologies, buffers are now commercially available, which allow faster and efficient reaction.

For ligation reaction in cloning experiments, we have two types of DNA molecules, the vector and the insert. During ligation, there are several possibilities of ligation product depending on the nature of the molecules and their concentrations. For example, the vectors or inserts may self-ligate giving circular monomers, may form linear dimers or higher multimers, and may form circular dimers or higher multimers and so on (Plate 2).

The desired recombinant product may be obtained by carefully controlling the relative concentrations. At low concentration, as the termini of different molecules have little opportunity to come close to each other, intramolecular interactions are favoured so that there is a tendency to form self-monomers. Intermolecular interactions, on the other hand, are favoured at high DNA concentrations. If the concentration of the vector is increased without increasing the concentration of the insert, self-ligated vectors will be predominant. On the other hand, if insert concentration is increased enormously, insert–insert dimers will be formed. Optimal insert vector ratios have to be worked out. Sometimes it is done through a preliminary ligation experiment using different relative concentrations of insert and vector. It is generally predicted that insert vector molar ratios of 2 to 6 are optimum for single inserts. A further complication will arise while working with multiple inserts or a heterogeneous mixture of inserts as is required during the creation of a DNA library. Using more inserts will lead to ligation of multiple inserts into the vector. To obtain an efficient ligation reaction, apart from optimization of concentration, other strategies have been designed involving modifications of the vectors or inserts, as discussed in subsequent sections.

How to Control Self-ligation of Vector DNA?

In cloning, self-ligation of the vector DNA is a problem, and this is taken care of using different strategies, either by using alkaline phosphatase or by using double digestion or modifying the vector or insert.

Dephosphorylation of Vector

During the ligation process, the enzyme ligase requires the presence of a 5'-phosphate at the nicked site in one and a 3'-hydroxyl group in the other. If 5'-phosphate groups at the end of the vector DNA can be removed, the chance of recircularization of the vector molecules is reduced. Alkaline phosphatase is an enzyme that can remove a phosphate group. The most commonly used enzyme is calf-intestinal phosphatase (CIP). Before ligation, the vector DNA molecules are treated with CIP, resulting in dephosphorylated vectors so that self-ligation is prevented. Since the foreign DNA inserts still contain the 5'-phosphate groups, ligation of the vector to insert is possible. The process is described in Plate 3. The disadvantages are that CIP treatment often overdigest and modify the sticky ends. Also by using the above strategy, insert dimers or insert multimers cannot be prevented.

Double Digestion

A way to reduce self-ligation is to cut the vector with more than one enzyme, usually with two different enzymes. The vectors used nowadays have multiple cloning sites (for example, pUC vectors) where different enzyme-cutting sites are present within a small DNA fragment. The double digestion to prevent self-ligation of vectors is illustrated in Plate 4.

When the vector is digested with two different restriction enzymes, for example, *Bam*HI and *Pst*I, the small fragment existing between these two sites will be removed. As a result, re-ligation of the vector will be prevented. Now digesting the insert DNA with the same two enzymes will in effect make all colonies recombinant. The difficulty of using double digest lies in the choice of enzymes. Conditions have to be determined at which two enzymes will work efficiently. Sometimes digestions are done sequentially rather than simultaneously.

Restriction Fragment End Modification

It is most convenient to ligate the restriction fragments possessing sticky ends. But the limitation is that the ligation is very specific; only fragments with compatible ends can ligate. For example, fragments digested with *Bam*HI will not ligate with fragments generated by *Eco*RI and vice versa. Another way is to choose blunt-ended fragments. But the choice of enzymes in such cases is not much, resulting in the limitation of this strategy as well. Some other strategies could be modifications of the restriction site as described as follows.

Conversion of Sticky Ends to Blunt Ends by Trimming and Filling

A strategy employs the conversion of sticky ends to blunt ends. This can be done either by trimming the overhang, thus getting rid of the unpaired sequences, or by filling in the complementary sequences of the unpaired overhang to form the double strand. When the restriction fragment possesses a 5'-overhang like the one created by *Eco*RI, the 3'-OH group will present a primer site to be extended by DNA polymerase. Usually, Klenow fragment of *E. coli* DNA polymerase I is used for this purpose as it lacks 5'-3' exonuclease activity. The presence of exonuclease activity in the polymerase enzyme will remove the 5'-overhang and may remove many more sequences. Trimming is required for restriction fragments with 3' sticky ends, such as those generated by digestion with *Pst*I. T4 DNA polymerase, which has a 3'-5' exonuclease activity, can be used for this purpose. The conversion from sticky ends to blunt ends is illustrated in Plate 5.

Other methods include the following.
* Attaching linkers and adapters at the ends of fragments to make the insert and vector compatible for joining.
* Homopolymer tailing at the ends of fragments.

These strategies are discussed in detail in chapter 13.

PROPAGATION OF PLASMIDS CARRYING FOREIGN DNA MOLECULES

Now that we have seen how to construct hybrid DNA molecules, the next step will be to get these molecules into living cells so that the DNA can be replicated and the genes that the foreign DNA molecule codes can be expressed. This is achieved by transforming the plasmid DNA into a suitable host. All commonly used plasmid vectors, used to clone foreign DNA fragments, allow for the insertion of a single vector molecule into the host cell. This single molecule may be amplified many times within the host, but all of the resulting molecules are identical. As a consequence, if a mixed population of DNA fragments is ligated into a common vector and transformed into, say, *E. coli*, the resulting bacterial colonies will each contain one (and only one) type of rDNA molecule. The mixed population of DNA fragments is segregated into individual components during the transformation and cell growth processes. Two methods of transformation are in use: the chemical method, which allows plasmids to enter bacterial cells at varying efficiencies, and the electroporation method, which uses high-voltage electric pulses and has increased transformation efficiencies. Treating cells with calcium ions can make them competent for the uptake of DNA. The DNA may adhere to the surface of the cell, and the uptake is mediated by a pulsed heat shock.

Chemical Transformation

Since *E. coli* cells are not naturally transformable, chemical methods are required to induce competence or the ability to take up DNA. In general, competence is induced by the treatment of growing cells with calcium chloride. The role of Ca ions in the competence is not clearly known, but there are several postulates based on experimental evidences. The cell envelope of *E. coli* is composed of an outer membrane and an inner membrane, with the periplasmic space in between. The structure of the outer membrane can be conceived as a fluid mosaic model consisting of lipopolysaccharides, phospholipids, and proteins embedded in between. A number of zones of adhesion are found, which are formed by the fusion of the outer membrane with the inner membrane. Previous studies indicate that DNA molecules can be transported through the adhesion zones. Since the DNA molecule is negatively charged, it is repelled by the negative charges present in the outer membrane. The addition of Ca ions neutralizes this effect and favours the transport of DNA molecules. In addition, Ca ions can induce non-bilayer structures in total lipids and can also enhance the phase transition of phosphatidylglycerol and lipopolysaccharide. Overall, Ca ions reduce the stability of *E. coli* membrane and facilitate DNA uptake.

The recombinant plasmid is then incubated with the competent host cells on ice and subsequently subjected to a brief heat shock at 42°C. The brief heat shock alters the fluidity of the membrane and allows DNA to enter the cell through the zone of adhesion. A nutrient medium is added to the cells to help them grow for a single generation and allow the phenotypic properties conferred by the plasmid (for example, antibiotic resistance) to be expressed. Finally, the cells are plated out onto a selective medium such that only the cells that have taken up the foreign DNA will grow (Figure 3).

The molecular processes by which transformation occurs are not well understood. The improvements of transformation efficiency are done by some empirical experimentation. Transformation efficiencies can be increased in the following ways.

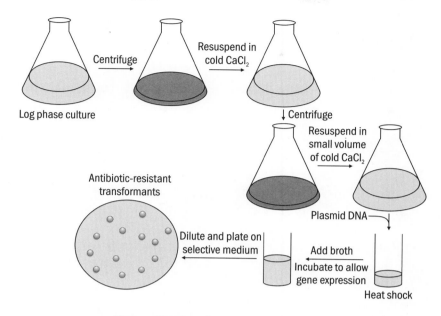

Figure 3 Transformation procedure

❖ By using bacterial cells deficient in restriction systems.
❖ By using bacterial cells mutated in certain exonucleases (for example, *rec*BC in *E. coli*).
❖ By using a mixture of Ca^{+2} and a variety of other divalent cations (for example, rubidium and manganese).

With the chemical transformation method, transformation frequencies of the order of 106 to 109 transformants can be obtained per microgram of plasmid DNA.

Electroporation

Electroporation uses an electric field pulse to destabilize the membranes of *E. coli* and induce the formation of transient pores within a biological membrane through which DNA molecules pass. This method was originally devised to introduce DNA into eukaryotic cells but has been subsequently adapted to transform *E. coli* cells by plasmids. The mechanism of electroporation is not well understood, and like chemical transformation, the procedure for a particular application is empirical. The parameters that influence the process are the types of host cells, the strength of the electric field, the amplitude and duration of the electric pulse, the concentration of plasmid DNA, and the composition of the electroporation buffer. Transformation efficiencies of 10^{10} transformants per microgram of DNA can be achieved by this method.

The advantages of electroporation are the reproducibility and high efficiency of transformation. The disadvantages are that in some cases, transformants generated by electroporation are marker dependent. For example, if a plasmid containing tetracycline and ampicillin resistance markers is introduced by electroporation, the number of tetracycline-resistant transformants is 1/100th of ampicillin-resistant transformants.

Advantages

❖ Electroporation is quite versatile, that is, effective with nearly all types of species.

❖ The efficiency is high as a large percentage of cells take in the target DNA or molecule.

❖ The amount of starting DNA required is small compared to other methods.

Disadvantages

❖ If the pulse time is long or the intensity is high, the size of the pores may become too large or the pores may fail to close after membrane discharge, leading to damage of the cell membrane.

❖ There could be non-specific transport of materials inside the cell during electroporation resulting in an ionic imbalance that may lead to cell death.

SUMMARY

- In DNA cloning, the foreign DNA is inserted in vitro into a vector, which carries the DNA into the host cell for further replication. Both prokaryotic and eukaryotic hosts can be used for cloning, and the choice depends on the application.

- Vectors used for cloning in bacteria are plasmids or phages; for eukaryotes, various yeast plasmids and transposable elements can be used in addition.

- Plasmids are usually used as cloning vectors. The plasmid cloning vectors used for cloning in bacteria are mainly pBR322 and pUC. The vector pBR322 has two antibiotic resistance markers and a number of unique restriction sites to which foreign DNA can be inserted. Insertion of foreign DNA disrupts one of these markers so that the selection of recombinant molecules is easy. pUC vectors have an ampicillin resistance marker and a polylinker containing multiple cloning sites in it. Insertion of a foreign DNA fragment interrupts the expression of a partial beta-galactosidase gene and can be easily detected with a colour test.

- The vector molecule is cleaved by restriction enzymes, and the foreign DNA molecule is joined to it by another enzyme ligase.

- The recombinant molecule is then propagated to the host by means of transformation, which can be done by chemical methods or by electroporation.

REVIEW QUESTIONS

1. Why is it necessary for a plasmid vector, which can be used for molecular cloning in bacteria, to carry a gene that can provide drug resistance to its host bacterium?

2. Which two enzymes are needed to produce recombinant DNA?

3. State the steps for cloning a foreign DNA into a plasmid and inserting the plasmid into the bacterium sequentially.

4. Describe the process of cloning a DNA fragment into the *Eco*RI site of the vector pBR322.

5. What do you mean by MCS? Write its other name and mention its advantage.

6. Explain the role of the ligase enzyme in gene cloning. Which reaction is catalysed by DNA ligase?

7. Suppose you are trying to clone a DNA fragment in a plasmid. You have a DNA preparation cut with restriction enzyme X. The gene you wish to insert has sites on both ends for cutting with restriction enzyme Y. You have a plasmid with a single site for Y but not for X. What will be your strategy to clone such a gene?

8. You have a polyclonal antibody to a protein. You would like to clone the coding sequence for the protein. How would you go about it?

Properties of Restriction Endonucleases

After reading this chapter, the student will be able to:
- Learn about naming restriction enzymes
- Describe the various types of restriction enzymes
- Describe endonucleases producing compatible ends
- Explain recognition sequence
- Understand cleavage pattern
- Describe restriction mapping
- Discuss mapping by double digestion
- Discuss mapping by partial digestion

INTRODUCTION

To investigate the function of a gene, we need to study the gene in isolation. This could be achieved by cutting the specific piece of DNA and cloning it to a suitable foreign organism, where it can be replicated. The ability to cut DNA at specific sites became possible by the discovery of restriction enzymes.

Restriction enzymes are members of the class of nucleases. Endonucleases cleave nucleic acids at internal positions, while exonucleases progressively digest from the ends of the nucleic acid molecules.

The restriction enzymes (or restriction endonucleases) cleave double- or single-stranded DNA at specific recognition sequences called restriction sites. Restriction enzymes are produced by bacteria and archaea, which use the enzymes for defence against viruses and bacteriophages. Inside a bacterial host, foreign DNA is restricted or cut (restriction), so that the phage is unable to infect the bacteria. The restriction enzyme cuts DNA at an internal phosphodiester bond; for double helix, two incisions are made, one at each strand. Thus the viral DNA is *restricted* in the bacterial cell by the restriction enzyme, and the bacterial DNA is *modified* by the methylase and is provided protection from its own restriction enzyme.

NAMING RESTRICTION ENZYMES

Since the discovery of restriction enzymes, many different enzymes have been isolated. Due to the availability of a large number of enzymes, a uniform nomenclature system was required. A suitable system was proposed by Smith and Nathans, and a simplified version of this is still in use today.

The names of restriction enzymes are directly derived from the bacterial strains from which they were first isolated. The commercially available enzymes are generally from engineered *Escherichia coli* expressing them. The key features of the naming system (Table 1) are as follows.

❖ The first letter of the genus name and the first two letters of the species name generate a three-letter abbreviation, which forms the first three letters of the name of the restriction enzyme. This abbreviation is always written in italics.

❖ The three-letter abbreviation is followed by the first letter of the strain.

❖ When a particular host strain has several different restriction modification systems, Roman numericals are used to indicate the order in which they were isolated.

TYPES OF RESTRICTION ENZYMES

Restriction enzymes are classified biochemically into three types (types I, II, and III) based on enzyme complexity, enzyme cofactor requirements, nature of the target site, and the sequence and position of the DNA cleavage. The distinguishing criteria of these three types of restriction endonucleases are summarized in Table 2.

Table 1 Examples of restriction endonuclease nomenclature

Enzyme name	Enzyme source and isolation
*Eco*RI	*Escherichia coli* strain RY13, first enyzme isolated
*Pst*I	*Providencia stuartii*, first enzyme isolated
*Hind*II	*Haemophilus influenzae*, strain RD, second enzyme
*Hind*III	*Haemophilus influenzae* strain RD, third enzyme isolated
*Taq*I	*Thermus aquaticus*, first enzyme isolated
*Sma*I	*Serratia marcescens*, first enzyme
*Bam*HI	*Bacillus amyloliquefaciens*, strain H, first enzyme
*Hae*III	*Haemophilus aegyptius*, third enzyme

Table 2 Properties of different types of restriction endonucleases

Property	Type I	Type II	Type III
Restriction modification activities	Nuclease and methylase are mutually exclusive	Separate nuclease and methylase reaction	Simultaneous nuclease and methylase activity
Protein structure	Three subunits	Homodimer	Heterodimer
Recognition site	Bipartite, random, and asymmetric (~15 base pairs)	4–6 base-pair sequence, most cases palindromic	5–7 base-pair sequence, asymmetric
Cleavage site	Non-specific (random), >1000 base pairs from the recognition site	At, or close to, the recognition site	About 25 base-pair downstream of the recognition site
Cofactor requirement	ATP, Mg^{2+}, AdoMet	Mg^{2+}	Mg^{2+}, AdoMet
Enzymatic turnover	No	Yes	Yes
DNA translocation	Yes	No	No
Site of methylation	Same as recognition site	Same as recognition site	Same as recognition site
Utility for manipulation	Not useful, due to non-specific cleavage	Very useful	Not useful, due to simultaneous restriction–modification activity

Type I restriction enzymes, such as *Eco*K and *Eco*B, cut DNA at random locations at some distance (1000 base pairs or more) from the recognition site. The recognition site is bipartite and interrupted, being composed of two asymmetric parts—one with 3–4 nucleotides and the other with 4–5 nucleotides, separated by a spacer sequence of 6–8 nucleotides. Type I enzyme is a multimeric complex consisting of restriction (*Hsd*R), methylation (*Hsd*M), and specificity (*Hsd*S) subunits. The cofactors required for cleavage are S-adenosyl methionine (AdoMet), hydrolysed adenosine triphosphate (ATP), and magnesium (Mg^{2+}).

Type III restriction enzymes, such as *Eco*P15, cleave DNA approximately 25 bases from the recognition site. Recognition sites are non-palindromic, usually two separate non-palindromes oriented inversely. Type III enzymes consist of restriction and methylation-specificity subunits. The cofactors required for cleavage are non-hydrolysed ATP, AdoMet, and Mg^{2+}.

Type II restriction enzymes are the most useful commercially available enzymes. These enzymes recognize and cleave DNA at the same site. The recognition sites are usually palindromic, but occasionally, partially palindromic or interrupted palindromic. Type II enzymes are homodimeric, and Mg^{2+} is required as a cofactor. Other divalent cations like Mn^{2+}, Cu^{2+}, Co^{2+} or Zn^{2+} when used may give rise to star activity (Box 1). Type II restriction enzymes generate two different types of cuts, depending on the position in the recognition sequence where the enzyme cuts. "Blunt ends" are generated if the enzyme cuts at the centre of the recognition site in each strand, while cohesive or sticky ends are generated if it cuts asymmetrically at the recognition sequence in two strands. The resulting fragment DNA will then have an overhang at each strand.

Box 1   Star activity

If the reaction conditions for restriction digestion are not met appropriately, the cleavage is not specific. The enzymes may cleave sequences that are similar but not exactly identical to the recognition sequence. This altered specificity is often called "star activity". Some of the conditions that may contribute to star activity are (1) higher concentration of enzyme compared to the DNA present, (2) presence of high concentration of glycerol (restriction enzymes are stored in glycerol), (3) use of non-optimal buffer, (4) presence of organic solvents, and (5) use of divalent cations other than Mg^{2+}.

Depending on the deviations based on the characteristics of type II enzymes, a number of subdivisions have been made, which are as follows.

❖ Type IIB restriction enzymes (for example, *Bcg*I and *Bpl*I) have recognition sites that are bipartite and interrupted. They cleave DNA on both strands; hence, both sides of the recognition site are excised. Both AdoMet and Mg^{2+} are required as cofactors.

❖ Type IIE restriction enzymes (for example, *Nae*I and *Nar*I) require two copies of the recognition site for cleavage. The enzyme binds to these two sites but only one acts as the target and is cleaved while the other serves as the allosteric effector and improves the efficiency of cleavage.

❖ Type IIF restriction enzymes (for example, *Ngo*MIV), like type IIE, bind to both the recognition sites but, unlike type IIE, cleave coordinately at both the binding sites.

❖ Type IIG restriction enzymes (for example, *Eco*57I) have restriction and modification activities in the same subunit but require the cofactor AdoMet to be active.

❖ Type IIM restriction endonucleases (for example, *Dpn*I) are able to recognize and cut methylated DNA.

❖ Type IIP enzymes recognize symmetric sequences.

❖ Type IIS enzymes (for example, *Fok*I) recognize non-palindromic asymmetric sequences and cleave at least one strand outside of the recognition sequence. These restriction enzymes exist primarily in monomeric form. Only Mg^{2+} is required as cofactor.

❖ Type IIT restriction enzymes (for example, *Bpu*10I and *Bsl*I) are composed of two different subunits. Some recognize palindromic sequences, while others have asymmetric recognition sites.

Isoschizomers

Restriction enzymes with the same recognition sequences are called isoschizomers (Greek *iso* = equal; *skhizo* = to split). However, isoschizomers may or may not cleave DNA identically to produce the same ends. The first enzyme to recognize and cut a sequence is termed the prototype, and the subsequent enzymes found recognizing the same sequence are isochizomers of the prototype. For example, *Hpa*II (C^CGG) and *Msp*I (C^CGG); *Sph*I (CGTAC^G) and *Bbu*I (CGTAC^G) are isochizomers, which have the same recognition site and cut at the same place within the recognition sequence. Other isoschizomers that recognize the same sequence but cut at different positions within the recognition sequence are *Acc*651 (G^GTACC) and *Kpn*I (GGTAC^C). They generate different sticky ends.

Neoschizomers

Neoschizomers are enzymes that recognize the same sequence but cleave at a different position than their prototype. Neoschizomers are a subset of isoschizomers. Examples are *Aat*II (GACG^TC) and *Zra*I (GAC^GTC); *Sma*I (GGG^CCC) and *Xma*I (G^GGCCC).

Isocaudomers

An enzyme that recognizes slightly different sequences but produces the same ends is an isocaudomer. In some cases, only one out of a pair of isoschizomers can recognize both, the methylated as well as unmethylated forms of restriction sites. In contrast, the other restriction enzyme can only recognize the unmethylated form of the restriction site. This property of some isoschizomers allows the identification of the methylation state of the restriction site while isolating it from a bacterial strain. For example, the restriction enzymes *Hpa*II and *Msp*I are isoschizomers as they both recognize the sequence 5'-CCGG-3' when it is unmethylated. But when the second C of the sequence is methylated, only *Msp*I can recognize both the forms while *Hpa*II cannot.

ENDONUCLEASES PRODUCING COMPATIBLE ENDS

Two or more restriction endonucleases recognizing identical or different sequences may generate identical DNA fragment termini. These endonucleases are said to produce compatible ends. DNA fragments with compatible termini may be ligated with DNA ligases to produce hybrid DNA molecules. For example, *Bgl*II recognizes a different six-nucleotide sequence than *Bam*HI but generates cohesive termini compatible with those of *Bam*HI. The hybrid sites generated by joining *Bam*HI and *Bgl*II cohesive ends cannot be cleaved by either enzyme.

*Bgl*II cleavage:	*Bam*HI cleavage:
5' A^G-A-T-C-T 3'	5' G^G-A-T-C-C 3'
3' T-C-T-A-G^A 5'	3' C-C-T-A-G^G 5'

RECOGNITION SEQUENCE

A recognition sequence is a site on a DNA sequence where it is cut by a restriction enzyme. Type II restriction enzymes recognize sequences 4–8 base-pair long. Hexacutters (6-base-pair-long recognition sites) are most commonly used in recombinant DNA technology and molecular biology. Many of the type II restriction endonucleases recognize palindromic sequences. Palindromes are words that read the same from left to right or right to left, such as "radar" and "madam". A restriction site palindrome is one that reads the same in the top (5' to 3') and bottom (complementary) (5' to 3') strand (Figure 1).

Some examples of common restriction endonucleases and their source and recognition sequences are presented in Table 3.

CLEAVAGE PATTERN

Type II enzymes exhibit strict sequence specificity. For example, a single base-pair change in the recognition site eliminates the enzyme activity. Two types of cleavage patterns are generally observed. Enzymes that make a staggered cut in the two DNA strands generate a single-stranded "tail" in both the cleaved fragments. The region known as "sticky ends" can transiently base pair with other DNA fragments containing complementary sticky ends, such as *Eco*RI and *Bam*HI. Other enzymes cut both the strands of DNA at the same position within the restriction site and generate "blunt ends", which means the entire fragment is double stranded, such as *Alu*I and *Sma*I.

RESTRICTION MAPPING

Restriction mapping describes the relative positions of restriction endonuclease cleavage sites within a linear or circular DNA. If two independently isolated fragments of DNA show a region of common restriction sites, it is expected that these two fragments have

Figure 1 Reading a palindromic sequence

Table 3 Source, recognition sequence, and end produced by some common restriction endonucleases

Restriction enzyme	Source	Recognition sequence	Ends produced
EcoRl	Escherichia coli	Uncut	Sticky
HindIII	Haemophilus influenzae	5′-G-A-A-T-T-C-3′ 3′-C-T-T-A-A-G-5′	Sticky
HaeIII	Haemophilus aegyptius	5′-G-G-C-C-3′ 3′-C-C-G-G-5′	Blunt
HpaII	Haemophilus parainfluenzae	5′-C-C-G-G-3′ 3′-G-G-C-C-5′	Sticky
Pstl	Providencia stuartii	5′-C-T-G-C-A-G-3′ 3′-G-A-C-G-T-C-5′	Sticky
Smal	Serratia marcescens	5′-C-C-C-G-G-G-3′ 3′-G-G-G-C-C-C-5′	Blunt
BgIl	Bacillus globiggi	5′-A-G-A-T-C-T-3′ 3′-T-C-T-A-G-A-5′	Sticky
BamHI	Curtobacterium albidum	5′-G-G-A-T-C-C-3′ 3′-C-C-T-A-G-G-5′	Sticky
Alul	Arthrobacter luteus	5′-A-G-C-T-3′ 3′-T-C-G-A-5′	Blunt

overlapping regions. Presence of closely related repeat sequences in two fragments will also show common restriction sites. Restriction mapping is the first step in characterizing an unknown DNA, especially in the absence of sequence information and is also a prerequisite for DNA manipulation like subcloning. Subcloning is the cloning of fragments of recombinant DNA into another vector. Knowledge of unique restriction sites on the DNA fragment can be used for subcloning.

Restriction mapping involves breaking of the segment of DNA into pieces with restriction endonucleases. The fragments are then separated in an agarose gel. The sizes of the fragments generated are then determined with the help of a DNA marker run on the same gel. To identify the relative locations of the enzyme sites on the segment of the DNA, two strategies are used, which are as follows.

1. Double digestion involving complete digestion of the DNA fragment with one enzyme at a time, followed by digestion with a combination of enzymes.
2. Partial digestion where digestion is controlled in such a way that not each cleavage site is cut. The two strategies are explained in the examples given in the subsequent sections.

MAPPING BY DOUBLE DIGESTION

An example of mapping by double digestion is illustrated in Figure 2 where a 7 kb DNA fragment to be mapped is considered. The fragment is at first separately digested with *Hind*III and *Pst*I. Digestion with *Hind*III yields two fragments of sizes 6.2 kb and 0.8 kb, while digestion with *Pst*I yields fragments of sizes 5.8 kb and 1.2 kb. It is apparent that the unknown DNA fragment has one *Pst*I and one *Hind*III site. To determine the location of the restriction site, double digestion was performed when the unknown DNA fragment was digested with *Hind*III and *Pst*I together. This double digestion produced three fragments of sizes 5.8 kb, 0.8 kb, and 0.4 kb, respectively. Here the 5.8 kb fragment is the same as that obtained with *Pst*I digestion alone. The 0.8 kb and the 0.4 kb fragments suggest that *Hind*III cuts within the 1.2 kb fragments of *Pst*I. Since the 0.8 kb fragment is generated by digestion with *Hind*III alone, the location of *Hind*III can be easily determined and is as shown in Figure 2.

MAPPING BY PARTIAL DIGESTION

This procedure is illustrated in Figure 3. A 3 kb unknown DNA fragment cloned in a plasmid is first digested with *Eco*RI to linearize the plasmid DNA. It is then labelled with a radioisotope on only one end. Then by a second digestion with *Not*I, the 3 kb fragment is dissociated from the plasmid but remains labelled at one end. It is then subjected to partial digestion by *Pst*I enzyme to generate a map of *Pst*I in this unknown DNA fragment. Partial digestion is performed by using very small amounts of enzyme for short periods of time. The digestion products are run on an agarose gel. Due to end labelling, each product defines the distance of restriction endonuclease sites from the labelled end, thus creating a map for *Pst*I on the unknown fragment.

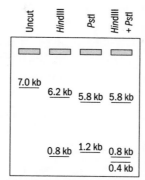

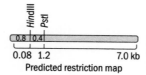

The restriction fragments generated by digestion with *Hind*III, *Pst*I, and *Hind*III + *Pst*I and separated by gel electrophoresis

Figure 2 Restriction mapping by complete digestion

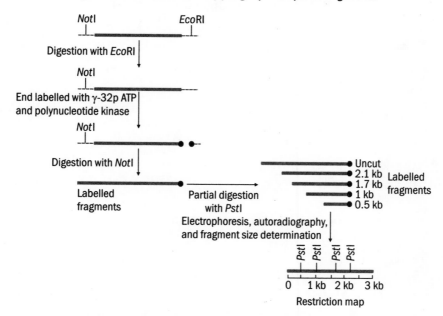

Figure 3 Restriction mapping by partial digestion

SUMMARY

- Restriction enzymes are endonucleases that recognize specific sequences on a DNA molecule called restriction sites and cut in both the strands. Restriction enzymes are produced by archaea and bacteria, which use these enzymes for defence against viruses and bacteriophages.

- Restriction enzymes are classified biochemically into three types, called types I, II, and III, based on enzyme complexity, cofactor requirements, nature of the target site, and the position of the DNA cleavage.

- Among these, type II enzymes are largely used in the molecular cloning of DNA. Type II enzymes typically cut DNA at specific 4 to 8 palindromic sequences producing defined fragments, depending on the position in the recognition sequence where the enzyme cuts. If the enzyme cuts asymmetrically at the recognition sequence in two strands, cohesive or sticky ends are generated. Blunt ends are generated when the enzyme cuts at the centre of recognition site on both strands.

- Restriction enzymes find useful applications in mapping genes and genomes. Restriction mapping describes the relative positions of restriction endonuclease cleavage sites within a linear or circular DNA. Any DNA can be mapped by using the double digestion method or partial digestion method.

REVISION QUESTIONS

1. What are restriction enzymes? Give two examples of restriction enzymes with their microbial source and restriction site.
2. Explain why the restriction enzymes were key to starting the genetic engineering revolution? Show how restriction enzymes produce sticky and blunt end cuts. Draw a typical restriction enzyme site and explain what a palindromic site is.
3. Determine the length and nature (5' or 3') of the overhang (if any) created by the following restriction endonucleases: *Eco*RI, *Hin*dIII, *Sma*I, *Bgl*II, *Bam*HI, *Alu*I, and *Hae*III.
4. What are the three types of DNA ends that can be generated after cutting DNA with restriction enzymes?
5. Find the restriction enzyme sites in these DNA sequences: AATCCTAGGACG; CCTAGT; AATCCTAGGACG; AAATTAATCGG; AAGGCGCGCCTAAT; TTGCATGCCTGCAGGTCGACTCTAGAGTATCCCCGGGTACCGAGCTC GAATTCACT.
6. What would happen to a bacterial cell if it contains no restriction enzymes?

7. How does a restriction enzyme act when it is used in constructing hybrid molecules of certain gene sequences and plasmid DNA?

8. What is the genetic function of restriction enzymes?

9. How does a bacterial cell protect its own DNA from restriction enzymes?

10. Do different restriction endonucleases ever generate sticky ends that are the same? Explain your answer.

4

Screening of Recombinant DNA Molecules

OBJECTIVES

After reading this chapter, the student will be able to:

- Discuss identification of recombinant plasmids by α-complementation
- Explain identification of recombinant plasmids by hybridization
  Discuss screening by polymerase chain reaction

INTRODUCTION

This chapter discusses ways to distinguish between bacteria containing recombinant plasmids and those containing empty vectors. It is rarely possible by visual inspection alone to find out which bacterial colonies contain recombinant plasmids. Sometimes, colonies containing recombinant plasmids are smaller. This is possible if the plasmid expresses a foreign protein that could retard the growth of the host cell, but the plasmids normally used for cloning do not express the foreign protein at a high level. In any case, the reduced size of the colony does not confirm the presence of a recombinant plasmid.

Another way is to use a plasmid having two selectable antibiotic markers (for example, tet^r and amp^r). If the foreign DNA is introduced with the disruption of one marker (say tet^r), the bacteria containing recombinant plasmids will be resistant to ampicillin but sensitive to tetracycline, while the bacteria with the empty vector will be resistant to both ampicillin and tetracycline. Over many years, several other methods have been devised. Here we will discuss three methods being widely used for selection in a molecular biology laboratory. These are as follows.

❖ Selection by α-complementation (blue-white selection)

❖ Colony hybridization

❖ Polymerase chain reaction (PCR) amplification of the cloned product

IDENTIFYING RECOMBINANT PLASMIDS BY ALPHA (α)-COMPLEMENTATION

Identification of foreign DNA inserts in recombinant plasmids has been made easier by the use of alpha (α)-complementation, which is nothing but a positive selection procedure for assessing whether a transformed bacterial colony contains plasmid with insert or not.

The *Escherichia coli lacZ* gene product beta (β)-galactosidase in its active form is a tetramer. Each monomer consists of two parts—an alpha fragment (N-terminal) and an omega fragment (C-terminal). If the alpha fragment is deleted, the omega fragment is non-functional and the tetramer formation is prevented leading to the loss of β-galactosidase activity. For example, in *E. coli del*M15 (DM15) strain, there is a small deletion in the gene coding for α-peptide and the strain is negative for β-galactosidase activity. However, if the alpha fragment expressed from a plasmid is introduced in *E. coli* DM15, the activity of the enzyme β-galactosidase can be functionally restored. This phenomenon is called α-complementation and has been used in vectors for selection. The cloning vector to be used for this purpose contains the *lacZ'* gene encoding the α-peptide of β-galactosidase as the marker (examples are pUC series, pBluescript). The multiple cloning sites (MCS, a short stretch of DNA with a number of closely spaced recognition sequences for common restriction endonucleases) are embedded in a functionally non-essential region of *lacZ'*. If the vector is successfully transformed into the *E. coli* DM15 host, it will produce α-peptides, and through intra-allelic complementation, functional β-galactosidase will be produced. Insertion of foreign DNA into the MCS will disrupt the *lacZ'* gene and will inactivate the α-peptide, and hence, β-galactosidase activity will be abolished. To detect α-complementation, the bacterial cells are plated onto a medium containing isopropyl thiogalactoside (IPTG, which inactivates the lac

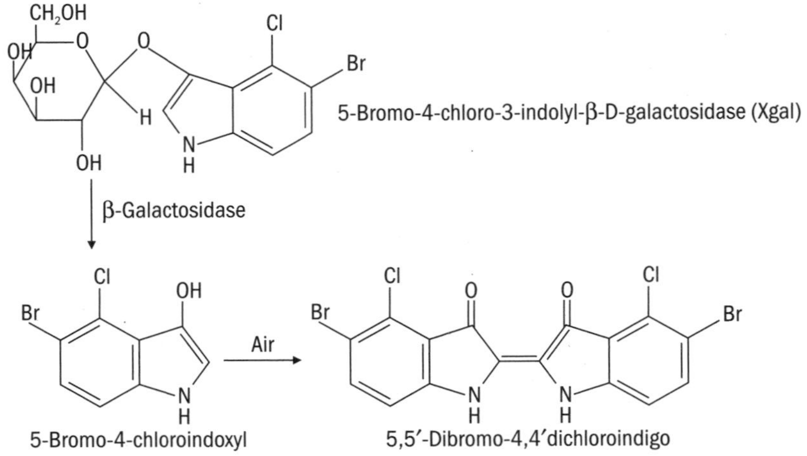

Figure 1 Cleavage of X-gal by β-galactosidase

repressor and serves as an inducer of β-galctosidase) and the dye 5-bromo-4-chloro-3-indolyl-β-D-galactoside (X-gal). X-gal is colourless, but on cleavage with the enzyme, β-galactosidase yields a derivative, which is blue in colour (Figure 1).

Plasmids containing intact *lacZ'* gene will produce β-galactosidase, which will give rise to blue-coloured colonies, but plasmids containing inserts will produce defective α-peptide and non-functional β-galactosidase and produce white colonies. The development of this colour test has greatly simplified the selection of recombinants. The structure of the recombinant can be further verified by plasmid DNA isolation and restriction digestion.

IDENTIFYING RECOMBINANT PLASMIDS BY HYBRIDIZATION

When a double-stranded DNA is heated or treated with alkali, the hydrogen bonds and other non-covalent forces holding the two strands of the molecule are weakened. As a result, two strands separate, and this phenomenon is called denaturation. When the conditions are reversed (the DNA solution is allowed to cool or neutralize with the acid), hydrogen bonds are reformed and the process is now termed renaturation and the double-stranded DNA fragment is annealed.

This principle has been utilized for the identification of foreign DNA fragments in a clone by a procedure called nucleic acid hybridization (Figure 2).

In this procedure, a single strand of DNA from one organism is bound to a nitrocellulose membrane (the membrane that binds single-stranded DNA only). Now a single-stranded DNA containing DNA sequences of the foreign DNA, but labelled radioactively, is added to the membrane. If the labelled DNA finds a complementary sequence, it will bind to it and a radioactive signal will be observed when the membrane is exposed to an X-ray film.

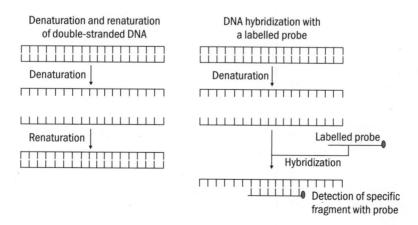

Figure 2 Denaturation, renaturation, and hybridization of DNA

Essentially, in this technique, bacterial colonies grown on agar plates are transferred to nitrocellulose membranes, lysed there, and the foreign DNA is identified by hybridization with a suitably labelled probe.

A sterile membrane filter is placed on the bacterial colonies to transfer them to the filter, which is then gently removed and prepared for hybridization. The bacterial colonies on these filters, which are treated with alkali to lyse them in situ as well as to denature DNA, and the single-stranded DNA are then bound to the membrane. Following neutralization, the sheet is treated with protease to remove proteins and leave the denatured DNA bound to the membrane. Finally, the sheet is baked at 80°C or treated with UV light to make the DNA firmly adhere to the membrane. The denatured (single-stranded) DNA can then be used as a template for the binding of complementary, radiolabelled DNA sequences. A suitably designed and labelled nucleic acid probe is then used for hybridization. The binding of these sequences to the membrane can be analysed by exposing the washed membrane to the X-ray film. The colonies containing identical, or at least similar, DNA sequences will produce signals.

The probe can be designed in four ways, which are as follows.

1. If the sequence of the cloned DNA is known, an oligonucleotide may be synthesized and used as a probe.
2. If a cloned DNA fragment of the same gene from a closely related species is available, this can be used as a probe.
3. If the complete sequence information of the gene of interest is available from the database, a polymerase chain reaction (PCR) can be used to amplify the gene in question to serve as a probe.
4. If the protein product is available, N-terminal sequencing of a few amino acids is done, and a nucleotide probe is designed following back translation from amino acids to nucleic acids using degenerate codons (Figure 3). The purity of the probe is important. If it is contaminated with vector DNA, it will hybridize with both recombinant and non-recombinant colonies.

SCREENING BY POLYMERASE CHAIN REACTION

PCR is an easy method of selecting recombinant clones. It is also possible to carry out what is called "colony PCR". Two oligonucleotide primers encompassing the target (here the inserted foreign DNA), or a part of it, are used for the specific amplification of the foreign DNA. Individual colonies to be tested are first lysed by boiling and then used in a normal PCR, where the required fragment, if present, will be amplified and yield a product. To monitor whether the PCR is working, a gene from the plasmid DNA is also amplified. The colonies that generate a product of appropriate size are then grown and re-screened for confirmation. This method can, nowadays, be used in a 96-well plate format, so that a large number of colonies can be screened at a time.

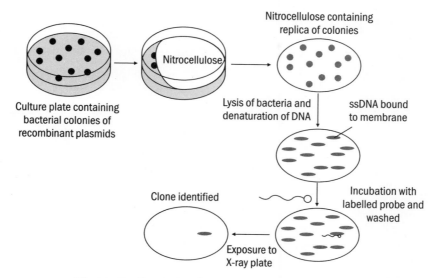

Figure 3   Screening by nucleic acid hybridization

Advantages of the PCR method over the hybridization technique are that PCR can be done with ease and is less time consuming. Moreover, there is no need to use radioactivity. Disadvantages are that one may pick up false positives by performing colony PCR; so confirmation of the product is needed, which can be done by sequencing a part of the product.

SUMMARY

- Screening of recombinant DNA molecules means distinguishing recombinant plasmid containing bacteria from those containing empty vectors. Cloning vectors like pBR322, which has two antibiotic resistance genes and a variety of unique restriction sites, is useful as one can introduce foreign DNA into a site that interrupts one of the antibiotic resistance genes.

- Screening of specific clones can be done by using α-complementation of β-galactosidase. Some vectors like those of pUC series have an antibiotic resistance gene and a multiple cloning site containing a number of restriction sites that interrupt a partial β-galactosidase gene, the product of which can be easily detected by a colour reaction. The recombinant clones have antibiotic resistance and cannot make active β-galactosidase enzymes.

- Recombinant DNA molecules can be identified by nucleic acid hybridization procedure. The bacterial colonies suspected to contain the recombinant plasmid grown on agar plates are transferred to nitrocellulose membranes, lysed there, and the foreign DNA is identified by hybridization with a suitably labelled probe.

- The probe may be an oligonucleotide homologous to the desired foreign DNA sequence, or a cloned DNA fragment of the same gene, or a PCR product from a closely related organism, or an oligonucleotide designed against the amino acid sequence of a gene product.

REVIEW QUESTIONS

1. Briefly explain a technique for visually screening transformed bacteria.
2. You have the cloned cDNA for a protein. Describe the procedures you would use to isolate a clone that contains the genomic sequence from a genomic library contained in lambda phage vectors.
3. Hybridization of nucleic acids is central to many molecular biological techniques. It relies on being able to label a DNA or RNA probe. List and describe two different methods for labelling nucleic acid probes.
4. You have cloned a DNA fragment in (a) the vector pBR322, (b) the vector pUC18. How would you screen for clones that contain an insert?
5. How can you use hybridization or expression to screen a library for a specific gene?
6. Which oligonucleotide primers could be synthesized as probes to screen a library for the gene encoding the peptide Met-Pro-Glu-Phe-Tyr?

Construction of DNA Library

INTRODUCTION

The collection of cloned deoxyribonucleic acid (DNA) fragments representing the genome (full or partial) of an organism is referred to as a DNA library. The collection of clones may consist of genomic DNA fragments or complementary DNA (cDNA) clones derived from a complex mRNA population. Accordingly, the library is referred to as a genomic DNA library or a cDNA library. This collection of fragments can be used to screen for specific functions or for other DNA sequences of interest.

For constructing such libraries, DNA fragments are generated by cutting DNA with a specific restriction enzyme and then ligated into a suitable vector. The collection of recombinant vector molecules is then transformed into host cells, one molecule in each cell. The total collection of all clones makes up the DNA library. To identify a target DNA, the entire library is screened with a molecular probe designed to bind with the target DNA. Once prepared, the library can be perpetuated indefinitely in the host cells and readily retrieved and searched for new functions as and when required. The two main types of libraries constructed and used are the genomic library and the cDNA library.

DNA libraries are useful to isolate or clone a single copy gene from a genome or a cDNA corresponding to an mRNA of interest. To achieve this, one could simply digest the total genomic DNA with a restriction endonuclease, insert all the fragments into a suitable cloning vector, and then attempt to identify the desired clone. Now the question that arises is how many recombinants should be screened for this? For example, the mammalian haploid genome contains approximately 3×10^6 kilo base pairs (kbp). If the fragment of interest is of 3 kbp, it will comprise only about one part in 10^6 of a preparation of genomic DNA. Therefore, about 1×10^6 independent recombinants must be prepared and screened to have a reasonable chance of including the desired sequence. Similarly, a rare mRNA species may comprise only one part in 10^5 or 10^6 of the total poly(A)-containing RNA. In either case, a large number of recombinants have to be included in the library. In theory, every DNA sequence in the target genome should be proportionally reflected in the recombinant DNA library. In practice, however, this is hard to achieve. A useful library, either genomic or cDNA, should have such a huge population of clones that it contains at least one version of every sequence of interest. The size of a library is thus dictated by the size of the cloned fragments and the size of the genome. The probability that a given DNA sequence is represented in a random library can be calculated from the following equation.

$$N = \ln (1 - P)/\ln (1 - f)$$

where N represents the number of independent clones in the library to be screened, P is the desired probability to isolate a particular sequence, and f is the fractional proportion of the genome in a single recombinant.

If I is the size of the average cloned fragment in base pairs and G is the size of the target genome in base pairs, then

$$f = I/G$$

For example, for a 99% chance ($P = 0.99$) of isolating an individual sequence of 20 kbp from a typical mammalian genome of 3×10^6 kbp,

$$N = \ln (1 - 0.99)/\ln [1 - (2 \times 10^4/3 \times 10^6) = 6.9 \times 10^5$$

This analysis assumes that the cloned DNA segments randomly represent the sequences present in the genome. In a strict sense, the randomness can be approached if the target DNA is cleaved completely in a random way prior to insertion into the vector. In practice, this randomness, too, is hard to achieve. In complex genomes, some of the sequences may be missing entirely, while others may be over-represented. Such a bias is inevitable as there is no method to generate a random fragment, especially for complex genomes.

Strictly, the randomness can be approached by mechanical shearing of DNA, which is a relatively inconvenient method. Normally, partial digestion is done with a restriction enzyme of choice to generate random fragments. The composition of the resulting fragments will depend on the distribution of the restriction sites within the genomic DNA and on the efficiency of digestion at different sites, which, in turn, is dependent on the neighbouring sequences. Also, regions rich in restriction sites seem to be under-represented because they may be reduced to unacceptably small sizes to be cloned. Regions with poor representation of restriction sites may be excluded as the size of the fragments may be too large to be accommodated in the cloning vector. Besides, there may be an additional bias as the library is replicated and expanded because some genomic clones may replicate better than others, and cloned sequences may undergo rearrangements during the passage. However, although the genomic library is imperfect, the proven performance of available libraries has made them time tested.

GENOMIC LIBRARY

A genomic library is one that contains DNA fragments representing the entire genome of an organism. It is constructed as follows and depicted in Figure 1.

❖ Genomic DNA is broken into manageable-sized pieces (for example, 15–20 kb for lambda vector and 40–50 kb for cosmid vectors) by physical breakage or by partial restriction endonuclease digestion. DNA fragmentation by physical means involves either vortexing or sonication. During sonication, ultrasonic pulses are sent at regular intervals through the DNA solution. In this way, DNA samples are hydrodynamically sheared. DNA solution is kept in ice since heat is evolved during the sonication process. The time of sonication can be varied to obtain DNA fragments of desired size. Any physical process will leave blunt-ended DNA fragments. In partial digestion, reaction conditions are standardized in such a way as to cleave any restriction site only occasionally. Thus all available sites in a DNA molecule are not cleaved. This generates a continuum of overlapping fragments. The advantages of restriction digestion are that cohesive ends can be generated to make ligation easy and the choice of enzymes is plenty so as to generate desired fragments. Disadvantages are that the size fractionation is non-random and many of the enzymes do not recognize methylated DNA.

❖ Optimal-sized fragments are then purified by gel electrophoresis or centrifugation techniques.

❖ The DNA fragments are inserted into a suitable cloning vector by the usual recombinant DNA techniques.

❖ The recombinant vectors are transformed to the respective host, where each host cell will take up one recombinant vector molecule. Thus each bacterial colony, selected after transformation, will contain a DNA fragment from the starting mixture. So we get a large collection of bacterial colonies, each carrying a foreign genomic DNA fragment.

CHOICE OF VECTORS

The plasmid vectors have the limitation that they cannot hold large DNA fragments (capacity ~2–10 kb). Moreover, the transformation of bacteria with plasmids is not very efficient. For the construction of DNA libraries, which often requires cloning of large fragments and a very efficient system of cloning, attention was drawn towards using other *Escherichia coli* vectors. The *E. coli* vectors, such as bacteriophage lambda or cosmid vectors, are typically used for the construction of genomic libraries.

Bacteriophage Lambda (λ) Vectors

Bacteriophage lambda (λ) is a temperate bacteriophage, which has two alternative life cycles—a lytic cycle and a lysogenic cycle. The wild-type bacteriophage λ virion contains the head, into which about 50 kb linear double-stranded DNA genome is packaged, and the tail with which it attaches on the surface of *E. coli* and infects it. When lambda infects *E. coli*, the linear DNA is injected into the bacterial cell where it is converted into a circular form due to the presence of *cos* sites. These *cos* sites, 12-nucleotide long, contain sequences complementary to one another and form base pairs with one another, thus creating a circular molecule. The DNA ligase within the cell then rapidly seals the

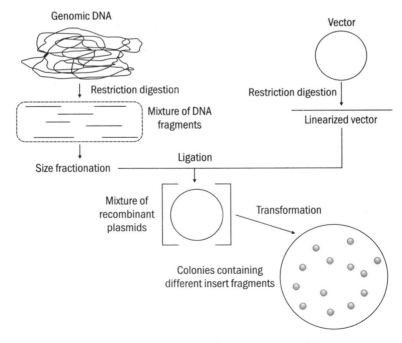

Figure 1   Construction of genomic DNA library

nicks in the circle to form a covalently closed circular DNA molecule. The lambda DNA then enters into one of the two alternative pathways—lytic or lysogenic (Plate 1).

The Lytic Cycle

The covalently closed DNA molecule initially replicates by theta mode, which is essentially similar to the replication of other circular molecules such as plasmids. At a later stage in infection, there is a switch from the theta mode to rolling circle replication, which yields a multiple length linear DNA molecule. The empty head and tail are synthesized at this stage. A protein attached to the phage head recognizes a *cos* site in the multiple length DNA molecule and initiates packaging of the DNA. This proceeds until the next *cos* site is reached, when the protein cuts the DNA at each *cos* site. These cuts are asymmetric, that is, the two strands are cut at positions that are not opposite to one another. There is a distance of 12 bases between the two cuts, leading to a sequence of 12 unpaired nucleotides at each end of the packaged linear DNA. The head is then sealed, the pre-formed tail is added, and the mature phage particles (about 100 progeny virions per cell) are released by the lysis of the cell.

The Lysogenic State

In the lysogenic state, the phage DNA molecule becomes integrated into the bacterial chromosomes, where it is passively replicated as part of the bacterial chromosome; phage particles are not produced. The inserted DNA is called a prophage, and the lytic cycle is prevented by the action of a specific repressor. The surviving cell is called a lysogen. If the lysogenic pathway is followed, then instead of replicating in the theta mode, the lambda DNA will integrate into the host chromosome by recombination, which occurs only at the specific attachment sites and is known as site-specific recombination. The site in the *E. coli* chromosome at which lambda integrates is located between the *gal* and *bio* operons. The attachment site is designated as *attB* since it is the attachment site on the bacterial chromosome, while the bacteriophage recombination site is designated as *attP*. The *attB* site consists of 30 base-pair sequence with a conserved central 15 base-pair region, where the recombination reaction takes place. The *attP* also contains identical core 15 bp region of *attB*. The flanking sequences on either side of *attP* are very important since they contain the binding sites for a number of other proteins, which are required for the recombination reaction.

Establishment of lytic or lysogenic cycle is controlled by two regulatory proteins, *cI* and *cro* gene products. The repressor CI can block the transcription of *cro* and vice versa. The relative expression of these two proteins determines the state of the lambda phage. If the lambda repressor CI dominates, the lambda genome integrates into the host chromosome and the lysogenic state follows. On the other hand, if the concentration of Cro protein is high, it turns off the *cI* gene, and the lytic cycle is initiated. The lysogenic state is favoured under normal growth conditions, but damage to host cells favours the lytic cycle.

The lambda map is illustrated in Figure 2. The genes show extensive clustering by function. The left half of the map consists entirely of the genes encoding head and tail proteins, and within this region, the head genes and the tail genes themselves form subclusters. The right half of the lambda genome shows gene clusters for DNA replication, recombination, and lysis. The genes are clustered not only by function but also according to the time during which their products are synthesized. For example, the N gene acts early; genes O and P are active later; and genes Q, S, R, and the head-tail cluster are expressed last. The transcription patterns for mRNA synthesis are thus very simple and efficient. There are only two rightward transcripts, and all late genes except for Q are transcribed into the same mRNA.

The potential of using the phage lambda as cloning vector was recognized in the late 1970s. Construction of a vector requires incorporation of foreign DNA of desired size into the vector and an efficient system to introduce the recombinant DNA into the bacteria. Since the genes of similar functions are clustered in the lambda phage genome, genes for lysogeny and other functions not required for lytic growth can be eliminated from the genome, and foreign DNA of similar size can be incorporated. The genome size required for efficient packaging into the lambda head is about 78%–105% of the wild-type genome so that about a maximum of 24 kb of foreign DNA can be inserted. Wild-type lambda DNA contains a number of target sites for most of the commonly used restriction endonucleases and is not itself suitable as a vector. Unwanted restriction sites have been eliminated, and a suitable polylinker has been introduced. Finally, the development of an in vitro packaging system has made lambda phage a powerful tool for cloning.

The central dispensable fragment of the lambda genome can be replaced by a fragment of heterologous DNA, leading to the construction of replacement vectors, such as Charon and EMBL3/4. Alternatively, small DNA fragments can be inserted without

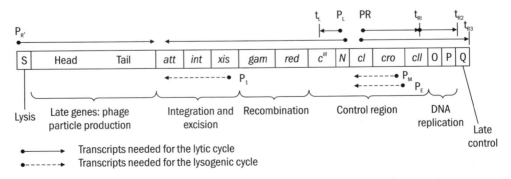

Figure 2 Lambda phage: genetic map and organization of transcripts

removing the dispensable region of the lambda genome, as in lambda gt10 and lambda gt11 vectors.

Derivatives of the wild-type phage have, therefore, been constructed and are mainly of two types.

1. Insertional vectors in which small foreign DNA fragments can be inserted into a specific restriction enzyme recognition site without removing the dispensable central segment of the lambda genome, for example, lambda gt10, lambda gt11 vectors.
2. Replacement vectors where the dispensable fragment (also called stuffer fragment) of the genome containing non-essential genes are replaced by the foreign DNA, for example, Charon, EMBL3/4.

The replacement vectors contain two target sites for restriction endonucleases. Cutting the DNA at these two sites will result in the production of three fragments—the left fragment, the right fragment, and the stuffer fragment. The stuffer, being dispensable, is separated and discarded. The left and right fragments, when joined, are too small to allow viable phage particles to form; they require an additional inserted piece of DNA for the formation of viable phage particles, which must be, in this case, 7–22 kb. Hence, not only does the cloning capacity increase, but such vectors also become useful for reasonably large inserts. Vectors of this type are known as replacement vectors (Figure 3).

Advantages of using lambda as cloning vector over plasmids are as follows.

❖ The size of foreign DNA to be inserted is higher for lambda phage compared to the insert size in plasmids.

❖ Infection process is a more efficient system for introduction of foreign DNA into bacteria compared to transformation used for plasmids.

❖ Phages multiply at a high copy number so that large amount of foreign DNA or proteins can be synthesized.

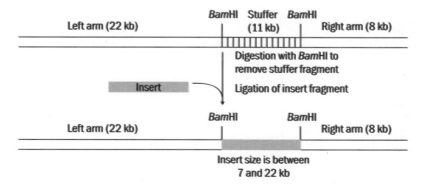

Figure 3 λ replacement vector

❖ If the cloned gene is toxic to the host, it does not matter as the toxin will not be synthesized until the phage infects the host, and following infection, the host will be killed anyway.

❖ Easy to store libraries in the relatively stable phage head.

Cosmid Vectors

In the normal life cycle of phage lambda, the phage DNA is replicated inside the host, following infection. The newly synthesized DNAs are then joined by their cohesive ends or *cos* sites to produce a concatemer (long chain of DNA molecule). During the packaging of the DNA into the lambda head, the *cos* site of the concatemer is recognized and cleaved enzymatically when each individual DNA molecule gets packaged into the phage head. Therefore, in principle, any plasmid DNA containing *cos* site can be packaged into the head of the phage lambda. Based on this principle, cosmid vectors have been constructed. Cosmids are essentially plasmid cloning vectors containing lambda *cos* site. The cosmid itself is too small to be packaged, but if the insert size is such that the total size of the cosmid becomes 78%–105% of the wild-type lambda genome size, it will be packaged easily to the lambda head. Therefore, cosmids enable large DNA fragments to be inserted. The structure of a cosmid vector is shown in Figure 4.

The packaging of cosmid DNA is done by in vitro packaging procedure. In this method, the plasmid containing the *cos* site is added to the lambda-infected cell extract, which contains heads and tails of the lambda phage. The cosmid DNA, by virtue of its *cos* site, will be taken up by the heads. In vitro, tails will be subsequently attached to the heads to make infectious particles that are finally introduced into bacteria. The success of cloning will be evident by the formation of phage plaques.

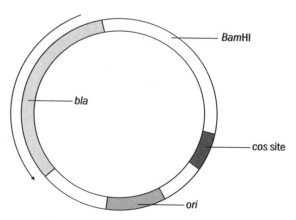

Figure 4 Cosmid vector

A major advantage of using cosmids is that the size of inserts is fairly uniform. The size of the insert is dictated by the DNA to be accommodated in the phage head. If the DNA inserted is too large, the two *cos* sites will be far apart, and the DNA will not fit into the phage head. On the other hand, if the insert is small, it will be unstable as the phage head will not be full. Thus the DNA cloned in cosmid will approximately be of similar size, which is sometimes important in making DNA libraries.

For many genome projects, it is essential to clone larger fragments of DNA. For cloning larger fragments, vectors such as P1-derived artificial chromosomes (PAC), based on the bacteriophage P1 (cloning capacity: 95 kb), or the so-called bacterial artificial chromosome (BAC) system, based on the F plasmid (maintains >300 kb), can be used. For still larger fragments, yeast artificial chromosome (YAC) vectors, maintained in yeast hosts, are often used and these typically carry inserts ranging from 0.3 Mb to 1.2 Mb of genomic DNA. Due to the size and complexity of YAC vectors, constructing YAC libraries requires special considerations for production, screening, and analysis. The alternative is to construct BAC libraries. Since large inserts are cloned in this technique, there is a greater chance of cloning a gene sequence with both, the coding sequence as well as the regulatory elements in a single clone.

Cosmids Versus Lambda

Cosmids and lambda vectors are preferred, one over the other, in different situations as follows.

❖ *Size of fragment*   Cosmids are preferred as they can carry larger fragments (~40 kb) compared to lambda vectors, which carry only about 20 kb.

❖ *Stability*   Since cosmid vectors are maintained as high-copy-number plasmids in *E. coli*, they have a tendency to be unstable, undergoing deletions that favour increased replication.

❖ *Screening*   Plaques (areas with no or little bacterial growth on a culture plate containing a lawn of bacteria) give less background hybridization than colonies; so screening libraries constructed in lambda phage give cleaner results compared to cosmid libraries by colony hybridization.

❖ *Storage and amplification*   Phage libraries can be stored almost indefinitely because phages have a long shelf life. The cosmid library obtained as bacterial colonies can also be stored, but bacterial populations cannot be stored as readily as phage populations, and there is often an unacceptable loss of viability.

cDNA LIBRARY

A cDNA library contains only cDNA molecules, which are synthesized from messenger RNA (mRNA). The mRNA population in a given cell represents only those genes that are expressed. Each cDNA clone in a library, therefore, contains the coding information of full or part of a gene.

In most eukaryotes, the genes contain introns interrupting the coding sequence (exons) or in the 5' or 3' untranslated regions, but these introns are rare in bacteria. The protein coding genes in eukaryotes are transcribed by RNA polymerase II. The resulting mRNA undergoes splicing to remove the introns cDNA so that the translation of a single contiguous message can occur. Further, a number of post-transcriptional modifications also occur, such as the addition of a 7-methylguanosine cap at the 5'-end and the addition of 100–200 adenine residues (a poly(A) tail) at the 3'-end of the transcript. Finally, the mature mRNA is transported to the cytoplasm.

The problem with mRNA is that it cannot be maintained in stable vectors and is difficult to manipulate. So a DNA copy (called cDNA) of the mRNA is required before a library can be constructed. cDNAs are created from the mature cellular mRNAs of eukaryotic cells by an enzyme reverse transcriptase.

Thus the size of the cDNA clone is significantly lower than that of the corresponding genomic clone in many cases. Since the cDNA library is representative of the RNA population from which it was derived, it will vary with the cell type or developmental stage from which the DNA has been isolated. The library will be enriched for abundant mRNAs but may contain a few clones of rare mRNAs. Also it will contain an alternative splice variant of the same gene.

The collection of cloned cDNAs has several utilities, which are as follows.

❖ cDNA libraries are used to express eukaryotic genes in bacteria either as a prerequisite for detecting the clone or because the primary objective is to express the polypeptide product. As bacteria do not have introns thus do not possess the enzymes required for splicing, eukaryotic cDNA clones find application where bacterial expression of the foreign DNA is necessary.

❖ Comparison of the cDNA sequence with the corresponding genomic DNA locus enables one to assign the position of intron/exon boundaries.

❖ The library is useful in reverse genetics. Unlike classical genetics (which begins from a phenotype, identifies the gene, and finally concludes with cloning and sequencing), the reverse genetics start with the cDNA library to find out the clones that are associated with a disease type.

❖ The cDNA library comes in handy for the isolation of the gene that codes for mRNA.

❖ It is easier to study cDNAs than mRNAs; cDNA clone collection can be utilized to determine the varieties of mRNAs present in a cell type or comparison of mRNAs between two cell types.

Procedure for cDNA Library Construction

A cDNA library can be constructed using the following steps.

❖ *Extraction and purification of mRNA*   The mRNA is extracted from cells and purified. Since mRNAs have a poly(A) tail, it is easy to distinguish and isolate them from the more abundant rRNA and tRNA molecules. All the RNA is extracted from cells and passed through a column containing oligo-dT-coated resins. RNA molecules that do not contain a poly(A) tail cannot adhere to such a column and will flow straight through it. The mRNA molecules, however, will bind to the column through complementary base pairing and will be eluted only when the concentration of salt flowing through the column is lowered (Figure 5).

❖ *First strand cDNA synthesis*   A cDNA copy of each mRNA molecule present is synthesized by mixing short (12–18 bases) oligonucleotides of dT with purified mRNA, reverse transcriptase, and four deoxynucleotide triphosphates (dNTPs) at neutral pH and in the presence of Mg^{2+} ions. The oligo dT will anneal to the poly(A) tail of the RNA molecule. Reverse transcriptase, an RNA-dependent DNA polymerase obtained from retroviruses like avian myeloblastosis virus (AMV) or moloney murine leukaemia virus (MMLV), uses the oligo-dT as a primer and synthesizes a single strand of cDNA on an RNA template. The resulting molecules will be double-stranded hybrids of one cDNA and one mRNA molecule. For cloning and other manipulation, the single-stranded cDNA has to be converted to double-stranded DNA, which is done with a DNA polymerase enzyme. This requires the removal of mRNA from cDNA–mRNA hybrid so that the cDNA can form a template for DNA polymerase to act upon. Removal of the RNA template is then achieved by treating the hybrids with alkali. As RNA is hydrolysed into ribonucleotides at about pH 11 and DNA is resistant to hydrolysis up to about pH 13, increasing the pH to about 12 results in the hydrolysis of the RNA, but not the DNA.

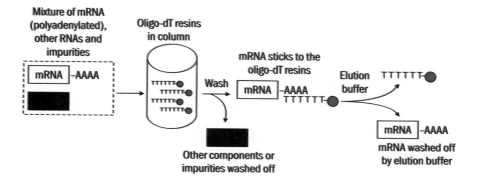

Figure 5   Principle of oligo-dT purification of mRNA

An oligo-dT primer used for making a cDNA strand will have heterologous ends. The primer can pair at numerous positions throughout the poly(A) tail and will consequently yield cDNA fragments of different lengths, even though they may have been derived from the same mRNA molecule. To overcome this problem, anchored oligo-dT primers are often employed. An anchored primer contains a G, A, or C residue at the 3'-end in addition to the 12–18 base dT sequence. Such primers will efficiently initiate DNA replication only if they are paired at the extreme 5'-end of the poly(A) tail, that is, only when the G, A, or C residue can base pair with the nucleotide immediately preceding the poly(A) sequence.

❖ *Synthesis of the second strand*   Using the first DNA strand as a template, the second strand DNA synthesis is carried out with a DNA-dependent DNA polymerase, such as *E. coli* DNA polymerase I. All DNA polymerases, whether they use RNA or DNA as the template, require a primer to initiate strand synthesis. However, beyond the poly(A) tail, mRNA molecules produced from different genes will be different. Therefore, a mechanism is required to initiate DNA synthesis at sequences corresponding to the 5'-end of the mRNA.

Early cDNA cloning strategies relied on the fact that the first cDNA strand has the tendency to transiently fold back on itself, forming a hairpin loop, which would serve as a self-priming structure for the formation of the second strand. The hairpin would be subsequently removed from the double-stranded cDNA by treatment with single-strand-specific S1 nuclease (Plate 2).

Since such methods invariably resulted in the loss of sequences at the 5'-end of genes, the second DNA strand is usually synthesized following either nick translation or homopolymer tailing. These methods also take care of the subsequent cloning step as they generate suitable ends for ligation to the vector DNA.

For nick translation, the RNA–DNA hybrid is partially digested with RNaseH, which recognizes the RNA component in an RNA–DNA hybrid and chew up RNA. Usually, it cleaves RNA non-specifically at a number of points to generate short oligonucleotides of RNA, which still remains attached to cDNA and can serve as a primer for fresh DNA synthesis to form double-stranded cDNA. Usually, the Klenow fragment of DNA polymerase I in the presence of the four dNTPs is used for this purpose. The Klenow fragment, often called large fragment of DNA polymerase I, is the 75 kDa fragment generated from DNA polymerase I by proteolysis with the enzyme subtilisin. This fragment is devoid of 5'–3' exonuclease activity but retains the 5'–3' polymerase and 3'–5' exonuclease activity of the parent enzyme. Commercially available Klenow is obtained by cloning and expresses a truncated version of pol A gene in *E. coli*. Absence of 5'–3' exonuclease activity is required to avoid the degradation of newly synthesized cDNA. The 3'–5' exonuclease activity, on the other hand, removes short RNA fragments attached to the single-stranded cDNA and is present in front of the enzyme. Finally, T4

DNA ligase is used to seal the remaining nicks (single-stranded DNA breaks) in the DNA backbone. The synthesis of double-stranded cDNA by this procedure is illustrated in Plate 3. The essence of nick translation is the simultaneous removal of DNA ahead of a nick and synthesis of DNA behind the nick. The net result is to move, or "translate", the nick in the 5'–3' direction. The resulting double-stranded cDNA molecule is blunt ended and can subsequently be cloned into a suitable vector, which is cut with a blunt-ended enzyme like *Sma*I.

Homopolymer tailing is an efficient way to clone cDNAs by using terminal transferase, a template-independent DNA polymerase that catalyses the addition of deoxynucleotides to the 3'-ends of DNA molecules. After first-strand synthesis, multiple C residues (homopolymer tail) are added to the 3'-ends by terminal transferase. An oligo-dG primer is then annealed to the oligo-dC tail, and the second strand is synthesized by DNA polymerase.

The next task is to clone the double-stranded cDNA to a vector. This is facilitated by a number of strategies, which include the use of linkers and adapter sequences to the ends of cDNAs.

Homopolymer Tailing

As mentioned, the oligo-dC end is added to the cDNA by terminal transferase in the presence of dCTP. The enzyme adds dCMPs, one at a time, to the 3'-ends of the cDNA. Thus with double-stranded cDNA, the enzyme adds poly(C) tails at the 3'-end of each strand. Terminal transferase uses either blunt-ended DNA or double-stranded DNA with 3'-overhang. For cloning, oligo-dG ends are added to the vector of choice. To add poly(G) tail, the vector is cut either with a blunt-end producing enzyme like *Sma*I or with *Pst*I, which produces a 3'-protruding end. The oligo-dC ends of the cDNA can then hybridize to the oligo-dG ends of the vector resulting in the insertion of the double-stranded cDNA into the vector creating a recombinant DNA. This process does not require any ligase, and the insert can be stably maintained and directly used for transformation. However, there may be gaps (not nicks) at the boundary of the vector and insert, which are repaired by the host enzymes following transformation of the recombinant plasmid. When the vector is cut with *Pst*I and poly(G) tail is added, it regenerates the *Pst*I site after annealing of the poly(C)-tailed cDNA. The insert can, therefore, be cut out of the vector using *Pst*I.

The homopolymer tailing method has the following advantages.

❖ Longer termini can be generated compared to the cohesive ends produced by a restriction enzyme, thus making cloning more efficient.

❖ Specificity is very high; the vector and insert have different complementary ends and the possibility of self-ligation is not there.

This method suffers from the following drawbacks.

❖ The addition of dNTP is often asynchronous, as different termini of DNA molecules carry homopolymeric tails of different lengths.

❖ The transformation efficiency of DNA molecules with homopolymer tailing is dependent on the genetic background of the host strain. This raises the possibility of systematic selection for or against a specific class of plasmid-cDNA hybrids (Plate 4).

Synthetic DNA Linkers

Synthetic linkers containing one or more restriction sites provide an effective method for joining double-stranded cDNA to both plasmid and bacteriophage lambda vectors and have superseded other methods of cloning cDNAs. Double-stranded cDNA molecules are incubated with T4 DNA polymerase or *E. coli* DNA polymerase I. The protruding single-stranded 3′ termini are removed by the 3′–5′ exonuclease activity of DNA polymerase, and the recessed 3′-OH termini are filled in by its polymerase activity. As a result, blunt-ended cDNA molecules are obtained. These molecules are incubated with a large excess of linker molecules in the presence of T4 DNA ligase, which catalyses the ligation of linkers to the cDNA molecules and results in cDNAs carrying linkers at the termini. These are then cleaved at a restriction site at the linker, purified, and ligated to a vector that has been digested with the same restriction enzyme (Plate 5).

Synthetic DNA Adapters

Adapters are short double-stranded oligonucleotides that carry one blunt end, which can be ligated to a double-stranded cDNA, and one cohesive terminus, which can be ligated to a compatible terminus in the vector. Adapters do not require digestion with restriction enzymes after they have been ligated to double-stranded cDNA. To avoid self-ligation, the termini of the cDNA molecules carrying adapters are phosphorylated with bacteriophage T4 polynucleotide kinase and then ligated to a large molar excess of the appropriate dephosphorylated vector.

VECTORS USED FOR CLONING cDNA

Till 1983, cDNA cloning was mainly done with plasmid vectors; so the library was in the form of bacterial colonies. The number of colonies would vary between experiments and is usually of the order of 10^5 colonies or more. These colonies can be stored at –70°C as glycerol stock (glycerol is used for cell stabilization) or can be maintained on nitrocellulose membrane. Preserving on nitrocellulose membrane is not very convenient as there is often a loss of viability, and the subsequent screening becomes laborious. The lambda phage libraries, on the other hand, can be stored indefinitely without the loss of viability. However, various lambda vectors are used for the construction of cDNA libraries, The characteristics are presented in Table 1.

Table 1 Vectors for cDNA cloning and expression

Vectors	Characteristics
Lambda gt10	Insertion vector. Can accept 7.6 kb of foreign DNA. The foreign DNA is introduced at a unique *Eco*RI cloning site, which interrupts the phage *cI* gene, allowing selection on the basis of plaque morphology. Libraries are screened by hybridization.
Lambda gt11	Insertion vector. Can accept 7.2 kb of foreign DNA. The foreign DNA is introduced at a unique *Eco*RI cloning site. Contains an *E. coli lacZ* gene driven by the *lac* promoter. cDNA can be detected by immunological screening, as well as by hybridization.
Lambda ZAP	A hybrid vector that possesses the attractive features of both bacteriophage lambda and plasmids. Often called phagemids. Up to 10 kb of foreign DNA can be cloned. A polylinker is present with six unique restriction sites, increasing cloning versatility and allows directional cloning. T3 and T7 RNA polymerase sites are available flanking the polylinker, allowing sense and antisense RNA to be prepared from the insert.
Lambda ZAPII	Equivalent to λZAP except that a Sam100 mutation has been removed, which allows better growth of bacteriophage λ, which, in turn, causes the plaques to turn blue much sooner in the presence of IPTG and X-gal.
Lambda ZAPExpress	Similar to other members of this series except that the excised plasmid carries control elements for expression of the cloned cDNA in eukaryotic cells. These elements include the powerful, immediate, early region promoter of the human cytomegalovirus positioned upstream of the multiple cloning site and an SV40 termination sequence located downstream from the cDNA insert. Excision of the expression plasmid from λZAP express is mediated by superinfection with a filamentous helper phage.
Lambda Ziplox	A hybrid of λgt11 that contains unique *Eco*RI, *Not*1, and *Sal*1 sites in the amino-terminal coding portion of *lacZ'* gene, encoding the α-complementation fragment of β-galactosidase. cDNA inserts up to 7 kb can be cloned. Screening is by nucleic acid or immunochemical methods. Cloned cDNAs can be excised on a replication-competent plasmid that is integrated into the vector DNA. Excision is accomplished by a *Cre/LoxP* mechanism in *E. coli* strains DH10B (ZIP), followed by plating on a media containing ampicillin.
Plasmids pSPORT1, pCMV-Script and pcDNA3.1	Incorporate a variety of sequence elements, including versatile multiple cloning regions, promoters for bacteriophage T3, T7, and/or SP6 polymerase, and a filamentous f1 phage intergenic region that permits DNA to be rescued in single-stranded form.

AMPLIFIED DNA LIBRARIES

The libraries that are freshly constructed are often called unamplified. However, in a strict sense they cannot be called unamplified because the transformed cells containing the clone grow into a colony, which is an amplification of the initial clone into a number copies equivalent to the number of bacterial cells present in a colony. Presently, for storage, individual colonies are spotted onto grids constructed on a suitable membrane or in microtiter plate wells (available in 96 well or 384 well formats). The membranes or microtiter plates are stored at –70°C in the presence of glycerol.

Amplified libraries are required in case of distribution of clones to different laboratories. The cells from primary source are then washed into culture medium, diluted, and glycerol is added for stabilization. Aliquots from the medium are then plated for the regeneration of the library. There is a possibility of distortion of the original library as different colonies may have unequal growth rate dictated by the nature of the insert.

SCREENING OF LIBRARIES

A DNA library consists of a million different clones and is built for studying gene functions, which involves searching for a specific recombinant clone in the library. The selection of a recombinant clone has been discussed in the previous chapters. The recombinant clones can be distinguished from non-recombinant ones using antibiotic resistance markers, blue-white selection, or by the size of the plasmid DNA. However, identification of the individual sequence or function of the recombinant portion of the clone requires some sort of specific selection process by which one molecule can be distinguished from another. The selection of an individual recombinant clone can be based either on its sequence or some function of the polypeptide encoded by that DNA sequence. Based on this, four methods are commonly available for screening DNA libraries (Figure 6).

1. Conventional nucleic acid hybridization (NAH), which requires a priori knowledge of the sequence or its homologues.
2. Amplification of the clone of interest by polymerase chain reaction.

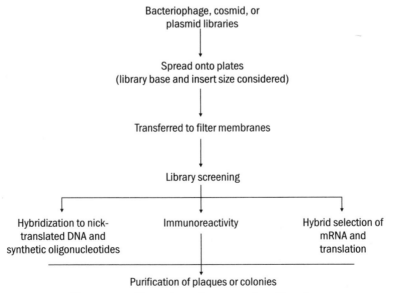

Figure 6 Flow chart for screening libraries

3. Interaction of the encoded polypeptide with the other factor (antigen–antibody interaction, protein–ligand interaction).
4. Any measurable biochemical function of the polypeptide by biochemical assay.

Nucleic Acid Hybridization

NAH (Box 1) is the most commonly used method of library screening because it is rapid, can be applied to large number of clones, and does not require an antigenically or biologically active product to be synthesized in the host cell. In case of cDNA libraries, NAH can be used to identify clones that are not full length (and, therefore, cannot be expressed). Furthermore, since the advent of the technique in the 1980s, it has been refined and studied, and the theoretical basis is well understood. This has led to the development of a large number of techniques with variations in the nucleic acid probe design.

This method relies on the hybridization of complementary single-stranded nucleic acid sequences (probes). For screening a library, colonies or phage plaques are transferred to a filter and subsequently hybridized with a labelled probe. The probe must contain at least part of the exact nucleic acid sequence of the desired genomic or cDNA clone. Rather than radioactive labels, a variety of non-isotopic labels are now commonly used. The position of a positive reaction on the filter can be correlated with the position of a specific clone, which can be picked off and purified. Various types of probes are being used with varying lengths and specificity, depending on the application.

Homologous Probes

A homologous probe is a hybridization probe that is exactly complementary to the nucleic acid sequence being searched. Homologous probes are used when at least a part of the sequence of the desired clone is known. For example, when a partial clone of an existing cDNA is used to isolate full-length clone from a cDNA library, this type of probe comes handy. Hybridization with homologous probes is carried out under stringent conditions.

Heterologous Probes

If it is known or suspected that the gene required is related to one that has already been characterized either from the same species or from another species, that DNA sequence can be used as a heterologous probe. Usually, a low stringency of hybridization (to allow for a degree of mismatching between the DNA sequences) is required.

If the probe does not produce reliable signals with DNAs of the target species in hybridization, it can be used in a zoo blot, which contains genomic DNA from a wide variety of species. For example, to search for a gene of yeast, the yeast probe may

Box 1 Nucleic acid hybridization

The nucleic acid hybridization (NAH) technique is based on the fact that two single-stranded nucleic acid fragments having complementary base sequence can rejoin to form a double-stranded hybrid molecule. This principle can be utilized to fish out a desired DNA fragment from a mixture of hundreds of DNA fragments. NAH requires two types of complementary single-stranded DNA molecules to be incubated under appropriate ionic strength and temperature to promote hybridization. Both types of reacting molecules may be present in solution, or one of them may be immobilized on a nitrocellulose membrane (filter).

A typical hybridization experiment is illustrated in Figure A. DNA is first digested with appropriate restriction enzyme and separated by gel electrophoresis. It is then denatured by alkali and then transferred to a nitrocellulose membrane and fixed on to it by heating at 80°C. The membrane is then incubated with radioactively labelled single-stranded DNA probe containing sequence complementary to one of the DNA fragments immobilized on nitrocellulose filter. The filter is then washed repeatedly to remove the unbound radioactivity, and the position of the bound probe is determined by autoradiography.

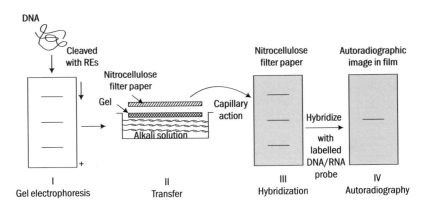

Figure A Experiment to illustrate hybridization

not cross-hybridize with blot containing genomic DNA from human, but may cross-hybridize with the blots containing genes from *C. elegans* or *Drosophila*.

Synthetic Oligonucleotide Probes

If an amount of purified protein is available, it is possible to determine the amino acid sequence of a small part of that protein (usually the N-terminal sequence). From this information, the likely sequence of the DNA can be deduced, and a short probe corresponding to that sequence can be synthesized and used for hybridization. However, due to the degeneracy of the genetic code, the same sequence of amino acids can be specified by many different oligonucleotides. To overcome these problems, the following strategies are commonly adopted.

❖ A family of oligonucleotides can be synthesized covering all possible sequences that can be derived from the amino acid sequence at hand. The oligonucleotides are mixed and hybridized with the idea that at least one of them should match and hybridize. To keep the size of each family within a manageable proportion, short oligonucleotides (14–17) of minimum size, practical for hybridization, are generally used.

❖ A longer oligonucleotide (40–60 base pairs) of unique sequence can be synthesized using the most commonly used codon for each amino acid. In that case, the exact match may not be possible but will give a signal under non-stringent conditions.

❖ An oligonucleotide that contains a base, such as inosine, at positions of high potential degeneracy can be synthesized. Inosine can pair with four conventional bases without seriously compromising the stability of the resulting hybrid.

Screening a DNA Library by Polymerase Chain Reaction

Polymerase chain reaction (PCR) may be used to screen genomic DNA or cDNA libraries for the identification of the gene of interest. For example, pools of plasmids or phagemids growing in individual wells of a 96-well plate can be lysed and screened by PCR using two oligonucleotide primers specific for the desired gene. Pools that generate a product of appropriate size are then grown and re-screened by the same procedure.

In another strategy, PCR primers are derived from the sequence of related genes in the data banks from the conserved region and used for PCR amplification from the species of interest. The PCR product can then be labelled and used to screen a library.

The use of PCR can often bypass the need to make a gene library at all. While studying a specific gene in closely related species, if the sequence of the gene from one species is known, the primer sequences from the known gene can be simply derived and used to amplify the corresponding gene from the other organisms.

Screening Gene Libraries with Antibodies

If antibodies directed against the protein of interest can be produced, these can be used to screen a DNA library in much the same way as described for a gene probe. Usually, nitrocellulose papers imprinted with lysed bacterial clones are soaked in a solution containing the antibody.

There are certain key factors for consideration. The gene of interest must be expressed to produce a protein; the antibody should recognize the protein in a denatured state, and it must recognize the primary product of translation rather than some modified form (for example, a glycosylated epitope) since this post-translational modification may not occur in the recombinant bacterium.

Screening is easier and more sensitive if polyclonal antibodies are used, but such antisera will also contain cross-reacting antibodies raising non-specific binding. Background or non-specific binding is much reduced by using monoclonal antibodies, which react only with a specific epitope, thus making the screening specific. Despite these limitations, antibody screening has been widely used, especially for genes coding for key antigens of specific pathogens.

Box 2   Functional screening of cDNA libraries for desired protein

Functional screening is a powerful procedure to identify a protein from millions of complementary DNA (cDNA) clones in a library. A typical procedure is illustrated schematically in Figure B. A cDNA library is divided into several sub-libraries. The pool of clones of each sub-library is then tested for a function by a suitable functional assay. If a positive response is obtained in a sub-library, each individual clone is tested to identify the desired one.

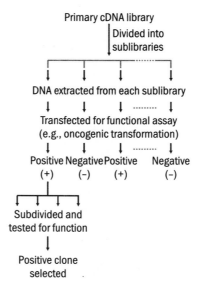

Figure B   Functional screening of cDNA libraries

Screening Gene Libraries by Complementation

It is also possible to screen a gene library by testing it for its ability to complement specific mutant strains of *E. coli*. For example, a leucine auxotroph (which is unable to grow without added leucine) may be converted to prototrophy by a recombinant plasmid carrying a *leu* gene from another bacterium. The required clone can, therefore, be identified by plating the library onto a medium lacking leucine. Of course, this procedure works only if the cloned gene is expressed and only if the product is functional (and readily detected) in *E. coli*. It is, therefore, not the most powerful way of screening gene libraries, but it is particularly useful for confirming the identity of cloned genes when a possible function (Box 2) has been identified by sequence comparisons.

The development of modern vectors and cloning strategies has simplified library construction to the point where many workers now prefer to create a new library for each screening, rather than risk using a previously amplified one. Furthermore, pre-made libraries are available from many commercial sources, and the same companies often offer custom library services. These libraries are often of high quality, and such services are becoming increasingly popular.

GENOMIC DNA LIBRARY VERSUS cDNA LIBRARY

The construction of genomic and cDNA libraries is represented schematically in Figure 7.

The essential differences between genomic and cDNA libraries are summarized in Table 2.

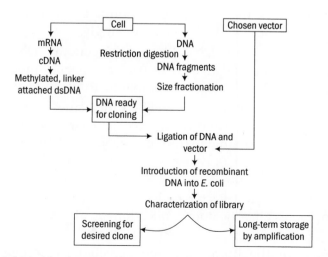

Figure 7 Schematic showing the construction of cDNA or genomic DNA libraries

Table 2 Comparison between genomic and cDNA libraries

Genomic DNA library	Complementary DNA library
A genomic DNA library is produced when the complete genome of a particular organism is cleaved into thousands of fragments, and all the fragments are cloned by insertion into a cloning vector.	A cDNA library is produced when all cDNAs synthesized from a mature mRNA template in a reaction are catalysed by the enzyme reverse transcriptase are cloned.
The genomic DNA library includes introns, exons, junk DNA, and so on.	A cDNA library is produced by reverse transcribing the mRNA of cells. This would form a library that consists of the DNA sequences of all the RNA transcribed from the cDNA, and excludes the introns.
The genomic DNA is digested by restriction enzymes or chopped by physical methods. Size fractionation is required before cloning.	cDNAs are not as heterogeneous as they come only from the coding region.
Provides a complete set of genetic information, including coding DNA, which leads to the expression of genetic traits and non-coding DNA.	A cDNA library provides some portion of the transcriptome of the organism.
Genomic DNA libraries provide detailed information about the organism, but are much more resource intensive to generate and maintain.	A cDNA library lacks the non-coding and regulatory elements found in genomic DNA. While the information in cDNA libraries is a powerful and useful tool since gene products are easily identified, the libraries lack information about enhancers, introns, and other regulatory elements found in a genomic DNA library.

SUMMARY

- A DNA library is a collection of cloned DNA fragments representing full or partial genome of an organism. It may consist of genomic DNA fragment or cDNA clones and is termed genomic or cDNA library accordingly.

- For the construction of genomic DNA library, the genomic DNA is digested with a suitable restriction enzyme and ligated to bacteriophage lambda or cosmid vector, which can hold larger DNA fragments compared to plasmid vectors.

- A cDNA library contains only the cloned cDNA molecules synthesized from mRNA. For cDNA cloning, mRNAs isolated from a particular tissue or developmental stage are reverse-transcribed into single-stranded cDNAs, which are then converted to double-stranded form by the polymerase enzyme, and finally ligated to lambda phage vector or cosmid vector.

- The collection of recombinant vector molecules is then transformed into host cells, one molecule in each cell. The total collection of all clones makes up the DNA library.

- A target DNA within the library can be detected by screening the clones by hybridization to a radiolabelled oligonucleotide probe, the sequence of which is complementary to a portion of the target DNA. Screening cDNA library for function can be done by functional complementation or by antibody screening.

REVISION QUESTIONS

1. What is a genomic library and what is its application?
2. Give the major advantages of using bacteriophage instead of conventional plasmids for the construction of libraries.
3. What does the term cDNA stand for? How is cDNA prepared? What features of cDNA make it a useful research tool? What are the limitations that restrict the usefulness of cDNA clones? Are cDNA preparations expected to contain cut sites for restriction endonucleases? Explain your answer.
4. How is a cDNA library constructed?
5. What are the differences between a genomic library and a cDNA library?
6. Molecular cloning can be used to isolate genomic DNA as well as cDNA clones. List and describe each type of vector system that can be used to clone either type of DNA.
7. Briefly explain a technique for visually screening transformed bacteria.
8. You have the cloned cDNA for a protein. Describe the procedures you would use to isolate a clone that contains the genomic sequence from a genomic library contained in lambda phage vectors.
9. Hybridization of nucleic acids is central to many molecular biological techniques. It relies on being able to label a DNA or RNA probe. List and describe two different methods for labelling nucleic acid probes.
10. You have cloned a DNA fragment in (a) the vector pBR322, (b) the vector pUC18. How would you screen for clones that contain an insert?
11. How can you use hybridization or expression to screen a library for a specific gene?
12. Which oligonucleotide primers could be synthesized as probes to screen a library for the gene encoding the peptide Met-Pro-Glu-Phe-Tyr?

6

Sequencing by Sanger's Method

OBJECTIVES

After reading this chapter, the student will be able to:
- Discuss chemical DNA sequencing method (Maxam–Gilbert)
- Explain the principle of Sanger's method
- Describe DNA polymerase and primer labelling
- Explain automated DNA sequencing
- Discuss capillary electrophoresis
- Analyse the advantages of automated DNA sequencing over manual DNA sequencing

INTRODUCTION

We are all aware that the human genome sequence has been completed. We cannot think of a world without nucleic acid and protein sequences. The story of sequencing began with protein sequencing in as early as 1955 when Frederick Sanger reported the sequence of a protein (insulin) and was awarded the Nobel Prize in chemistry in 1958. It was observed that the arrangements of amino acid residues in a protein or polypeptide molecule followed an order, which is defined but arbitrary. This order was unique for a particular protein. Although it was realized that the DNA base sequence would determine the amino acid sequences in a protein, no progress had been made on the sequencing of DNA prior to 1970s. Sequencing of DNA or determining the exact order of the four bases adenine (A), thymine (T), guanine (G), and cytosine (C) in a sample DNA fragment was difficult compared to protein sequencing for the following reasons.

❖ Unlike proteins, which come in smaller and manageable sizes, the chain length of DNA was much greater and was difficult to approach.

❖ The chemical nature of different DNA molecules was very similar, and therefore, difficult to separate from each other by chemical methods, whereas proteins could be readily purified.

❖ Unlike proteins that can be cleaved by proteases at a specific site to generate overlapping fragments, no base-specific DNAs were known at that time.

❖ The 20 amino acids had varying properties, which made their separation easy but the existence of only four bases made the sequencing more difficult.

In spite of the above difficulties, the first DNA sequence reported was that of the cohesive ends of the bacteriophage lambda DNA, which was only 12 bases long, by Ray Wu and Ellen Taylor, Cornell University, Ithaca, USA. Prior to this, the RNA sequencing was done and the methodology employed for DNA sequencing was derived from the RNA sequencing method, which was analogous to protein sequencing method as RNAs with base specificity were known.

The process was tedious and not applicable to sequencing longer fragments of DNA. In 1977, American molecular biologists Maxam and Gilbert developed a chemical method for DNA sequencing, which could determine the sequence of a DNA molecule up to 500 base pairs and was favoured for quite some time. An enzymatic method was discovered almost in parallel by English biochemist Sanger in 1977. The complete genome DNA sequence of the bacteriophage phi174 containing 5386 base pairs was reported by Sanger. Gilbert and Sanger are considered pioneers of DNA sequencing. In 1980, they were jointly awarded the Nobel Prize. Earlier, DNA sequencing was done manually. The products of a sequencing run were separated by gel electrophoresis followed by autoradiography. The autoradiographs were read personally by individual workers and interpreted. Later, enzymatic sequencing became the method of choice as it had the potential for large-scale application and was refined and automated. Today, sequencing is done in central facilities where a series of automated procedures are adopted. The results are fed into the computer, and the output is received in the form of a chromatogram along with the sequence. The automation of DNA sequencing technology has created a revolution. A large number of complete genome sequences are available for all three kingdoms of life and this enormous data is stored in databases. These databases are open resources, which can be accessed and analysed for further research.

The advances in the technology for DNA sequencing required development in several fields like molecular biology (improvements in DNA isolation, gel electrophoresis, polymerase chain reaction [PCR], cloning and sequencing strategies), instrumental analysis leading to the development of automated instruments, and computer science, leading to the creation of powerful computers (hardwares and softwares), which can handle huge data. A few selected applications of DNA sequencing are presented in Box 1.

CHEMICAL DNA SEQUENCING METHOD (MAXAM–GILBERT)

In the chemical DNA sequencing method innovated by Maxam and Gilbert, the DNA to be sequenced was radioactively labelled at one end, then cleaved chemically to produce

---

Box 1   Applications of DNA sequencing

DNA sequencing has now become an indispensable technique in a molecular biology laboratory. Its application is now at every step, be it gene cloning or genome analysis. Some of the key applications are as follows.

❖ Sequencing gene information is important for planning the procedure of cloning and further manipulation to study its function.

❖ Polymerase chain reaction (PCR) amplification of a DNA fragment or gene requires the knowledge of the flanking sequences of the DNA fragment.

❖ DNA sequencing helps to determine the restriction sites in a plasmid, which will be useful in cloning a foreign gene in a plasmid.

❖ In eukaryotes, sequencing can be used for the identification of exons and introns of a gene.

❖ Sequencing can be used to identify the site of mutation.

At the genomic level, sequence determination of a number of genomes has been effected. The complete genome sequence would lead to the following.

❖ Construction of restriction endonuclease map or physical mapping of genome.

❖ Open reading frame (ORF) searching or determining the presence of a gene.

❖ Determination of the presence of repeats, tandem repeats, inverted repeats, hairpin formation, and others.

❖ Phylogenetic studies and construction of molecular evolution map.

❖ Identification of genetic variation or polymorphism.

❖ Mapping of diseased gene in human genome.

---

a mixture of oligonucleotides in such a way that they differed by a single nucleotide at the other end. The cleavage was done at each of the four bases in four different reactions. This mixture of oligonucleotides was then separated by high resolution electrophoresis on polyacrylamide gels, and the position of the bands was determined by autoradiography.

The first step consists of end-labelling (5′-end) the double-stranded DNA molecule with radioactive phosphorus (^{32}P). The DNA strands are denatured and treated with alkaline phosphatase to remove the 5′-phosphate. It is then reacted with ^{32}P-labelled ATP in the presence of polynucleotide kinase, which attaches a labelled phosphate at the 5′-termini. The DNA fragment is then subjected to base-specific chemical treatment. Four different sets of chemical reactions are set up, which selectively modify the DNA backbone at *G*, *G + A*, *C + T*, or *C* residues. Each reaction involves the modification of the base, removal of the modified base from its sugar, and DNA strand cleavage in the sugar (Figure 1). The reactions are explained as follows.

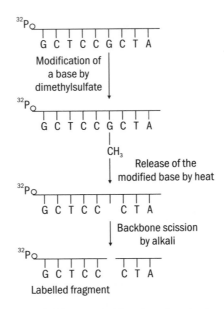

Figure 1 DNA sequencing by chemical method

❖ *G reaction*   Dimethyl sulphate is used to methylate guanine at N-7 position; the glycosidic bond of methylated *G* becomes unstable and is broken by heating at neutral pH. Further heating with alkali cleaves the sugar from the neighbouring phosphate groups.

❖ *A + G reaction*   Since dimethyl sulphate (DMS) also methylates adenine at N-3 position, the procedure is the same as that of *G* reaction, except that a dilute acid is added after the methylation step. The glycosidic bond of methylated adenosine is less stable than that of methylated guanosine; thus gentle treatment with dilute acid releases adenine preferentially. Subsequent alkali treatment will yield strong *A* and weak *G* pattern.

❖ *C + T reaction*   Hydrazine reacts with *C* and *T*, cleaving the base and leaving ribosylurea. Hydrazine may react to form hydrazone. After partial hydrazinolysis, the DNA is treated with piperidine to cleave the sugar-phosphate backbone.

❖ *C reaction*   The presence of high concentration of NaCl preferentially suppresses the reaction of thymines with hydrazine.

❖ The four bases are incubated with piperidine, which cleaves the sugar-phosphate backbone of DNA next to the residue that has been modified.

The base-specific reactions are designed such as to modify only a small proportion of the susceptible bases. Each reaction tube thus contains a mixture of single DNA strands of various lengths, each of which contains ^{32}P label at one end and chemical cleavage point at the other. These samples are then run in high resolution polyacrylamide gel in parallel lanes and separated electrophoretically. A specific band pattern is created, which can be directly read after autoradiography.

Chemical degradation method was initially popular, but only largely superseded by the enzymatic method. The disadvantages of the method are as follows.

❖ Chemical reactions are slow.

❖ The method requires the use of hazardous chemicals and, therefore, needs special handling.

❖ The method is difficult to automate and, therefore, is laborious and time consuming.

It is used only where it is required to sequence DNA directly rather than via enzymatic copy. It is suitable for applications like studies on DNA modification, secondary structure formation, and in some instances of protein interaction.

PRINCIPLE OF SANGER'S METHOD

Sanger's method is essentially an in vitro mimic of the in vivo DNA replication process. During in vivo replication, an RNA primer (a short RNA sequence that primes the DNA synthesis reaction) attaches to the region of DNA called the origin of replication. DNA synthesis is then catalysed by DNA polymerase in the presence of deoxynucleotide triphosphates (dNTPs), which are incorporated one after the other.

The bases join in a sequence, which is complementary to the template strand. Since DNA polymerase can elongate in 5′ to 3′ direction, one strand is continuously replicated while the other strand proceeds in pieces.

The DNA sequencing reaction starts with a double-stranded DNA fragment. For in vitro DNA synthesis, a small piece of DNA having sequence complementary to any strand of the template (5′–3′) serves as the primer. The DNA template is then denatured when the primer anneals to its appropriate position (3′–5′ complement). DNA polymerase then synthesizes DNA in 5′ to 3′ direction. This will continue till the bases are added up to the end of the template.

For DNA sequencing, Sanger utilized the above reaction with a modification. He had the idea of stopping the reaction prematurely after each base so that the base sequence can be read. This was done by modifying the nucleotide in such a way that once this modified nucleotide is incorporated, the next nucleotide will not be able to attach to it and the reaction will be terminated. Now an incoming nucleotide joins its previous nucleotide by forming a covalent bond between 3′-OH of the previous sugar and 5′-phosphate of the incoming base. If the 3′-OH of a nucleotide is replaced by H, the incoming nucleotide will be unable to join and the reaction will stop. A nucleotide modified in the above manner is called dideoxynucleotide triphosphate (Figure 2) and serves as a terminator.

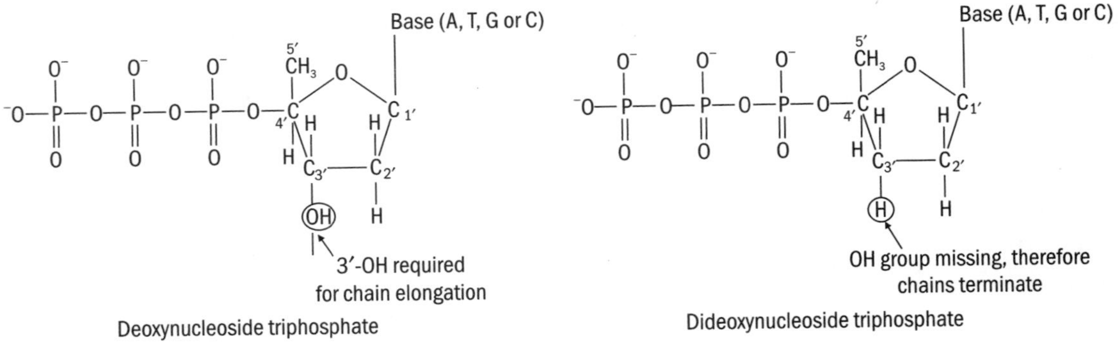

Figure 2   Structures of deoxynucleotide triphosphate and dideoxynucleotide triphosphate

PROCEDURE OF SANGER'S METHOD

For a given region of DNA to be sequenced in practice, the sequencing method consists of the following steps as illustrated in Figure 3.

First, a double-stranded DNA to be sequenced is taken, which serves as the template DNA or the original sequence. The DNA template is then mixed with a primer. The primer is a known sequence about 12–24 bases long and complementary to the 3'-end of one strand of the template DNA. This primer sequence is so called because it primes (or starts) the enzymatic reaction. The template DNA is next denatured by heat or alkali when the primer anneals to the template.

Optimal size of primers has to be chosen; too short primers will have the chance of finding complementary sequences at many places and may anneal, while too long primers may make annealing unstable. The primer is synthesized chemically, and annealing is effected through hydrogen bond formation between the primer and the complementary DNA strand. The primer is radioactively labelled at its 5'-end with ^{32}P-phosphate. The end labelling is done by the enzyme polynucleotide kinase and adenosine triphosphate (ATP) having gamma-phosphate labelled with ^{32}P. When the radiolabelled primer is annealed to the template DNA, each product formed following extension by DNA polymerase will have ^{32}P label at its 5'-end.

The solution is then divided into four tubes labelled as *G, A, T,* and *C* containing the following reagents.

G: all four dNTPs, ddGTP, DNA polymerase

A: all four dNTPs, ddATP, DNA polymerase

C: all four dNTPs, ddCTP, DNA polymerase

T: all four dNTPs, ddTTP, DNA polymerase

The DNA polymerase faithfully elongates a complementary copy of single-stranded DNA template. The dNTPs are added sequentially to the growing primer chain, and a base complementary to the base in the template strand is selected. A phosphodiester

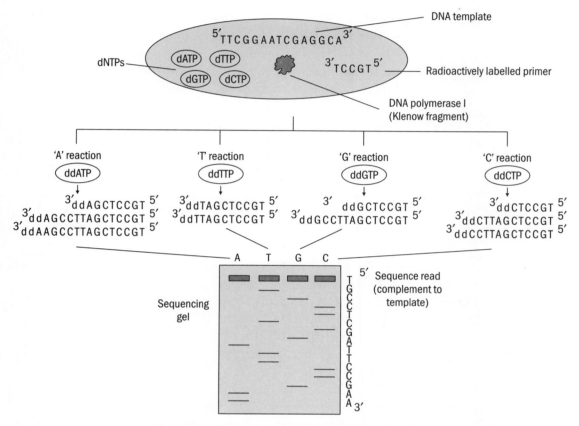

Figure 3 General strategy for DNA sequencing

bond forms between the 3'-hydroxyl group at the growing end of the primer and 5'-phosphate group of the incoming dNTP.

Whenever a ddNTP is incorporated, the phosphodiester bond cannot form and so the chain is terminated. These ddNTPs are devoid of an OH group at the 3' position of deoxyribose, instead an H is present. Absence of a 3'-OH will not allow the next nucleotide to be incorporated as the 5'-phosphate of the next nucleotide cannot form a bond with 3'-H.

In this method, all four reactions start from the same nucleotide but end in a specific base, which is different for each tube. In a solution containing the reactants, the same chain of DNA will be synthesized over and over again; but as soon as a ddNTP is incorporated, the newly synthesized chain will be terminated. As ddNTPs are randomly incorporated, synthesis terminates at many different positions at each reaction tube. Each of the ddNTPs is used at about 1% of the concentration of dNTPs. The DNA polymerase will integrate dNTP or ddNTP at random; so using the above concentration of ddNTP, the strand will continue to be synthesized in 99% cases but will be terminated at 1%. In this way, bands of all lengths will be produced.

After the completion of the reaction, the products (newly synthesized and labelled DNA fragments) are denatured by heat into single DNA molecules and subjected to denaturing urea polyacrylamide gel electrophoresis (PAGE). The electrophoresis is carried out at high voltage with positive electrode at the bottom and negative electrode at the top. The labelled fragments migrate from top to bottom according to their sizes, the shorter fragments move faster. Each reaction is loaded onto a separate lane. The resolution of gel electrophoresis should be such that DNA molecules of 50, 100, 150, or 200 bases can be differentiated from those having 51, 101, 151, and 201 bases. For better resolution, large gels are used because in large gels, the DNA molecules will move further, and separation between two consecutive bands will be greater. A high concentration of urea (6 to 8 M) has been used by various investigators to prevent the folding of molecules and secondary structure formation (which alters the mobility in a gel). The samples are denatured before loading, and the gels are run at higher temperature (50°C) to prevent hydrogen bond formation.

Following electrophoresis, the gel is dried on a chromatography paper and exposed to X-ray film. Drying of gel is necessary to reduce the thickness of the gel and prevent it from cracking. Also bands are less diffused if a dry sample is exposed to X-ray film rather than a wet one. The template strand, which is not labelled radioactively, will not produce any band on the X-ray film. The other strand producing radioactively labelled DNA fragments will appear as a dark band on the X-ray film. A dark band in a lane, therefore, represents a DNA fragment obtained as a result of chain termination due to the incorporation of a dideoxynucleotide. For each reaction, the chain termination event nearest to the primer will yield the smallest fragment, while the chain terminations far away from the primer will give larger fragments. In PAGE, the DNA strands are ordered by size; the smallest fragment will migrate to the bottom, while the larger fragment will move slowly and remain near the top of the gel. As each of the four reactions is loaded onto the gel in four consecutive lanes, the DNA sequence can be read from the bands forming ladder on the gel. The 5′ to 3′ sequence is obtained from the relative positions of the different bands among the four lanes, from bottom to top. It is to be noted that this sequence is complementary to the template strand that was initiated. This method is often termed the manual sequencing method as the sequence has to be read manually following gel electrophoresis and autoradiography.

MODIFICATIONS

For reliable and accurate DNA sequencing, some modifications ultimately culminated into the advent of automated DNA sequencing. The modifications included replacement of *Escherichia coli* DNA polymerase with DNA polymerase of high processivity or later with thermostable DNA polymerase. The technology of DNA labelling was also modified over the years in order to overcome the disadvantages of using ^{32}P labelling.

DNA Polymerase

In Sanger's method, the Klenow fragment of *E. coli* DNA polymerase I was used. The disadvantage of Klenow fragment is that it discriminates ddNTPs and, therefore, requires high concentration of ddNTPs. These ddNTPs are expensive and also, in the process, increase background noise.

Later, the Klenow fragment was replaced by "Sequenase", a natural or modified form of T7 DNA polymerase with higher processivity. Compared to the Klenow fragment, it can polymerize longer run of nucleotides before releasing them from the template. As a result, the incorporation of dideoxynucleotides is less affected by local nucleotide sequences, and the sequencing ladders comprise a series of bands with more even intensities. Moreover, the presence of tyrosine in the active site (phenylalanine in case of Klenow fragment) decreases the discrimination of ddNTPs.

Taq DNA polymerase, the DNA polymerase from the thermophilic bacterium *Thermus aquaticus*, has also been used. Being thermostable, it can be used in a chain-termination reaction carried out at high temperature (65–70°C), which minimizes the chain termination artifacts caused by the secondary structure in the DNA.

Primer Labelling

The labelling of primers by ^{32}P-phosphate has many advantages and disadvantages. As the half-life is small (14 days), DNA can be labelled to a very high specific activity. Specific activity measures the amount of radioactive disintegrations per unit concentration so that labelling with ^{32}P means that larger number of disintegrations are generated at low concentration of the product (primers). Disadvantages are that the radiolabelled substrates have a short shelf life. The signal is often foggy because the beta particles produced by ^{32}P, due to their penetrating power, generate a spread-out signal on the X-ray film. The ^{32}P labelling has been replaced by much lower energy ^{33}P, which forms a sharper image on the X-ray film; the half-life is twice as long as that of ^{32}P. A second solution is the use of ^{35}S thiophosphate (sulphur replaces one oxygen in the phosphate making it a thiophosphate). This has a longer shelf life (half life is 87 days) and a less energetic radioactive decay, which makes the bands sharper.

Non-isotopic methods using chemiluminescent biotin–streptavidin system have also been used. In this system, the oligonucleotide at the 5'-end of the primer is linked to biotin. The enzyme alkaline phosphatase is then bound to the oligonucleotide by a streptavidin conjugate. The enzyme catalyses a luminescent reaction, which can be detected by a photographic film.

However, the most dramatic advancement of this technique has been the use of fluorescent dyes, which was later developed into automated DNA sequencing technology. The basic biochemistry of Sanger's method remains the same. The dideoxy method has been made more efficient by the use of a general primer for routine sequencing (Box 2).

Box 2 Creating a general purpose primer

In this strategy, the single-stranded DNA phage M13 vector has been used. This phage is similar to phiX174 as both are packaged as single-stranded DNA, and both are replicated to double helices within the host. The double-stranded form or the replicating form can be subjected to engineering by standard methods, while the single-stranded form can be used for sequencing. J Messing and his colleagues introduced a polylinker containing a variety of restriction endonuclease sites (which will be available for cloning) in a bacterial gene (*lacZ*) inserted into M13. The gene codes for beta-galactosidase enzyme that breaks down an artificial substrate X-gal, which is normally colourless but becomes blue upon cleavage by the enzyme. Thus when functional *lacZ* gene is present, M13 gives blue plaques. If the *lacZ* gene is disrupted by a cloned insert, X-gal remains intact, and hence the plaques are colourless. (M13 does not lyse *E. coli* cells and, therefore, forms turbid sites and not true plaques).

If an oligonucleotide primer is synthesized complementary to a region of the phage DNA upstream of the cloning sites, dideoxy sequencing of the cloned DNA can be done easily. Single-stranded phage DNA containing a cloned insert is isolated and hybridized with the synthetic oligonucleotide. This operation creates the primer configuration for dideoxy sequencing of the cloned DNA. Any suitable cloned DNA fragments can be sequenced using this method.

AUTOMATED DNA SEQUENCING

With the various advancements in technology, including the tremendous power of computers, the original Sanger's method, which was largely manual, has become outdated. The most promising advancement of DNA sequencing was the introduction of automated DNA sequencing technology, which took DNA sequencing to a high throughput platform, the fruits of which are seen today in the form of complete human genome sequencing. However, this new technology, which has replaced the earlier manual procedure, is based on the same principles of Sanger's sequencing method. In 1986, Leroy Hood's group at the California Institute of Technology, USA, devised a DNA sequencing method in which radioactive labelling, autoradiography, and manual reading of bases were replaced by fluorescent labelling, laser-induced fluorescence signal detection, and reading of bases by computer. These modifications dramatically increased the speed and ease of DNA sequencing.

In this method, a sequencing primer was end labelled with a fluorescent dye. Four different fluorescent dyes emitting fluorescence of different colours were selected. The same primer was labelled with four different dyes generating four different end-labelled primer sets. Four separate synthesis reactions were set up with dNTPs, ddNTPs, and DNA polymerase, and each reaction had a different set of primer (Plate 1).

Upon completion, the reaction products were combined and loaded on one lane on a sequencing gel with laser detector, and a computer was attached to it for recording the data (Plate 2).

Thus different sequencing reactions can be loaded onto separate lanes, and at least eight sequences can be done in parallel using this approach. The sequencing gel was scanned with a laser beam near the bottom of the gel. When a fragment passes the detector, the fluorochrome dye with which it is labelled will be excited by a laser beam, and the resulting fluorescence signal is sensed and detected. Continuous recording of the fluorescence data was done by a powerful computer over the period of the entire run (about 12–13 h). The sequence was deduced from the order in which the dyes moved past the detector. The outputs were received in the form of a chromatogram, the coloured peaks (intensity of emission at a given wavelength) showing the position of the nucleotide in the sequence. Software programmes were developed that could transform the entire data in the form of an actual sequence.

This method was commercialized by Applied Biosystems, a US-based life technology company located at Foster City, California, in 1987, which revolutionized the process of large-scale sequencing, making the genome sequencing project a major success.

Since then, a number of other modifications have been introduced; some of the advancements such as dye terminator method and cycle sequencing (Box 3) are discussed as follows.

Dye Terminator Method

In this procedure, instead of labelling the primer, fluorescent-tagged terminators are used. Each of the four ddNTPS is labelled with a different fluorescent dye and, therefore, emits different colours when fired with the laser beam of the detector. The sequencing reaction is started in a single tube with all four labelled ddNTPs in it. Whenever a ddNTP is incorporated, the growing chain is stopped, but each chain will have a fluorescent label at its 3'-end. Upon completion of the reaction, the product is loaded onto one lane of a gel, as described above. As the DNA fragment passes the detector, each ddNTP will fluoresce with a different colour. The recording, as before, is done by a computer.

Box 3   Cycle sequencing

The cycle sequencing method (Plate 3) is a modification of dideoxy sequencing. A dideoxy sequencing reaction mixture (template DNA, unlabelled dNTPs, fluorescently labelled ddNTPs, primer and thermostable DNA polymerase) is subjected to a polymerase chain reaction (PCR). It undergoes repeated rounds of denaturation, annealing, and synthesis steps in a thermal cycler. In this way, linear amplification of the sequencing products occurs, thus requiring fewer templates DNA compared to the conventional dideoxy method of sequencing. The key difference from the normal PCR is that only a single primer is used in the cycle sequencing reaction; so the amplification of the product is not exponential, but linear. Fluorescent fragments are generated in the usual way by incorporating labelled ddNTPs such that all terminated fragments will contain a fluorescent label at their ends.

Choice of Fluorescent Dyes

Each of the four dyes ideally should exhibit a strong absorption at a common laser wavelength, while the emission is at distinctly different wavelengths. (Different fluorophores affect the mobility of DNA fragments to different extent; this can be minimized by using the above criteria for the selection of fluorescent dyes.)

CAPILLARY ELECTROPHORESIS

The use of capillary electrophoresis has enhanced the speed of sequencing. Increase in the applied electric field will cause an increase in the sequencing speed, but application of high electric field will also generate heat. The thickness of a slab gel is such that it cannot radiate heat sufficiently and as such there is a limitation on the amount of electric field that can be applied. In capillary electrophoresis, high purity fused silica capillaries are used, which are about 50 μm in diameter. Because of this small diameter, these capillaries can sustain high electric field. Thus rapid separation of DNA fragments is possible.

The use of capillary electrophoresis has overcome two main problems of gel electrophoresis and is currently being used in place of DNA sequencing gels. These are as follows.

❖ The applied voltage during electrophoresis cannot be increased above a certain limit.

❖ The products can be detected only after complete separation, that is, only after it has reached the end of the gel.

Increase in applied voltage is required in order to increase the speed of sequencing. This will cause heat generation. The thickness of a slab gel is such that it cannot radiate heat sufficiently, leading to the limitation of the amount of electric field that can be applied. In capillary electrophoresis, high purity fused silica capillaries are used, which are about 50 μm in diameter. Because of this small diameter, these capillaries have a high surface to volume ratio so that these can radiate heat readily and can sustain high electric field. Thus rapid separation of DNA fragments is possible. Also the migrating molecules are detected by the use of the laser system, which is sent near the beginning of the capillary through which the samples pass through and the fluorescent light emitted is detected on the other side of the capillary by a charge-coupled device (CCD) camera. CCD then converts the resulting information into a digital data format, and the graphical presentation of the chromatogram is observed through a computer screen. The capillary tubes are filled with suitable matrices that determine the resolution limit. Currently used matrices are linear polyacrylamide (LPA) or performance-optimized polymer 6 (POP6). Physical properties and pore size of these two matrices differ—LPA is viscous and POP6 is like jelly. Sample loading in capillary electrophoresis is by electrokinetic injection, which involves the transfer of charged ions directly onto the separation matrix inside the capillary.

The following are the advantages of capillary electrophoresis over slab gel.

❖ The loading of gels is much easier. No casting of gels is required; the matrix is injected into the capillary before loading.

❖ Sample can be loaded easily with less error.

❖ Faster movement of DNA ions can be achieved as the capillary can sustain high voltage.

❖ The difficulty of lane tracking in slab gel electrophoresis is overcome as the data is collected at the beginning of the electrophoresis.

ADVANTAGES OF AUTOMATED DNA SEQUENCING OVER MANUAL DNA SEQUENCING

The principal advantages of this method over the conventional manual method of DNA sequencing are as follows.

❖ Elimination of the use of radioactivity.

❖ Gel processing after electrophoresis and autoradiography is not required.

❖ The tedious manual reading of the sequencing gel is not required as the data is directly fed into the computer.

❖ The need to run several overlapping gels for a given set of reactions is eliminated.

❖ The method is extremely fast and can be robot mediated if required.

SUMMARY

- DNA sequencing refers to the methods for determining the exact order of the four bases adenine (A), thymine (T), guanine (G), and cytosine (C) in a sample of DNA fragment.

- In the chemical DNA sequencing method innovated by Maxam and Gilbert, the DNA to be sequenced is radioactively labelled at one end, then cleaved chemically to produce a mixture of oligonucleotides in such a way that they differ by a single nucleotide at the other end. The fragments are separated by gel electrophoresis and are visualized by autoradiography.

- In the enzymatic method of DNA synthesis innovated by Sanger, four DNA synthesis reactions are carried out in the presence of four different chains terminating dideoxynucleotide triphosphates and radioactively labelled primers. The fragments are separated by gel electrophoresis, and the sequence is inferred from autoradiograms.

- Sanger's sequencing method has been refined and automated by the use of fluorescent labels; DNA synthesis is carried out by polymerase chain reaction (PCR) and computerized detection systems and is implemented in large-scale sequencing projects. The use of capillary electrophoresis has enhanced the speed of sequencing.

REVIEW QUESTIONS

1. What is the role of dideoxyribonucleotide triphosphates in Sanger's sequencing method? Explain.
2. Why is the Klenow fragment of DNA polymerase I used during DNA sequencing?
3. What are the advantages of using Sequenase over Klenow in DNA sequencing reaction?
4. How all growing chains are prevented from being terminated at the same length?
5. What would happen if dideoxynucleotide triphosphate concentrations are increased in Sanger's sequencing reaction?
6. Why are four parallel electrophoretic lanes used in the standard method of Sanger's DNA sequencing? How have newer methods eliminated this requirement?

7

Protein Production in Bacteria

OBJECTIVES

After reading this chapter, the student will be able to:
- Describe the protein expression system
- Explain *E. coli* protein expression system
- Discuss the general characteristics of the expression vector used in *E. coli*
- Describe the optimization of expression levels in *E. coli* expression system
- Explain promoter sequences
- Describe optimization of other factors influencing protein expression
- List the strategies for improving protein solubility
- Learn about improving protein stability
- Understand reducing protein toxicity
- Explain in vitro expression using *E. coli* extracts
- Understand co-expression and its methods
- Explain protein purification

INTRODUCTION

Proteins or polypeptides have many uses, including various enzymes added to washing powders and hormones used in therapeutics. Production and purification of proteins and their biochemical and structural analysis have contributed largely to the field of immunology, cell biology, drug discovery, and medicinal chemistry. For some proteins, the extraction is easy, and the raw material (for example, microbial culture or plant) is easily available. For others, the source may be scarce, such as proteins from human sources having medicinal use. For example, the hormone somatotropin has been used to treat a medicinal condition called pituitary dwarfism. The normal growth of a child is affected in this medical condition due to less production of somatotropin by the pituitary gland, and the patients have to be treated by supplying the hormone from other sources. For a long time, the only source from which somatotropin had been isolated was the pituitary gland of a dead person. However, it had two drawbacks: one, limitation of supply, and two, risk of transmission of other diseases from the dead person.

After the advent of recombinant DNA technology, attention was drawn to the fact that bacteria or yeast, which is grown easily, can be used as a host for heterologous protein expression by cloning the gene and coding for protein of interest in these organisms. In this way, the directed synthesis of large amount of desired protein can be achieved. A therapeutic product synthesized by this technology will also be free from the possibility of transmission of associated diseases and is much safer than the one produced from natural source. Thus the production of recombinant proteins is the major driving force for the commercialization of various biotechnological products. A list of recombinant proteins having therapeutic use is presented in Table 1. There is large-scale implementation of protein expression in research. For example, the proteins (transcription factors) responsible for the regulation of cellular activities or factors responsible for cell–cell communication are produced in minor quantities. Previously, only some sensitive assay systems could have detected their presence, and it was impossible to study these proteins as sufficient quantities of the latter were not available. Protein expression technology had facilitated the production of those proteins in substantial amounts for further study so long as the genes could be identified and cloned.

PROTEIN EXPRESSION SYSTEMS

For successful protein expression, the expression host system will need the following characteristics.

❖ Rapid growth in cost-effective media
❖ Host must be non-pathogenic to humans (biosafe)
❖ Eukaryotic proteins should fold successfully
❖ Post-translational protein modifications should be possible

The expression system can be grouped into categories as described subsequently and summarized in Table 2. Choice of a system depends on the application and also on the solubility, functionality, speed, and quantity required.

Table 1 List of recombinant proteins available in the market and their therapeutic use

Recombinant protein	Therapeutic use	Marketed by	Product
Human growth hormone (rhGH)	Growth defect usually for stunted growth	Eli Lilly	Humatrope
Somatotropin	HIV-wasting syndrome	Serono	Serostim
Biosynthetic human insulin (BHI)	Diabetes mellitus	Eli Lilly Nordisk	Humulin Novolin
Factor VIII	Haemophilia A	Bayer	Kogenate
Factor IX	Haemophilia B	Wyeth	Benefix
Erythropoietin (EPO)	Anaemia	Amgen	Epogen
Granulocyte colony-stimulating factor (G-CSF)	Cancer	Amgen, California, USA	Neupogen
Alpha-galactosidase	Fabry disease	Genzyme	Fabrazyme
Alpha-glucosidase	Pompe disease	Genzyme	Myozyme
Tissue plasminogen activator (TPA)	Acute myocardial infraction	Genentech	Activase
N-acetylgalactosamine-4-sulphatase	Mucopolysaccharidosis VI	BioMarin Pharmaceutical	Naglazyme
DNase I	Cystic fibrosis		Pulmozyme
B-Glucocerebrosidase	Gaucher's disease	Genzyme	Cerezyme
Interferon-β	Relapsing multiple sclerosis	Biogen Idec Serono	Avonex Rebif
Anti-EGFR Monoclonal antibody	Metastatic colorectal cancer	Amgen	Vectibix
Ig-CTLA4 fusion	Rheumatoid arthritis	Bristol-Myers Squibb	Orencia
Luteinizing hormone	Infertility	Sereno	Luveris
Anti-VEGF monoclonal antibody	Metastatic colorectal cancer and lung cancer	Genentech	Avastin
Anti-IgE monoclonal antibody	Asthma	Genentech	Xolair
Anti-CD11a monoclonal antibody	Chronic psoriasis	Genentech	Raptiva
Follicle stimulating hormone	Infertility	Sereno NV Organon	Follistim Gonal-F
Anti-CD52 Monoclonal antibody	Chronic lymphocytic leukaemia	Genzyme, Bayer	Campath
Anti-HER2 Monoclonal antibody	Metastatic breast cancer	Genentech	Hercepin
Anti-CD20 monoclonal antibody	Non-Hodgkin's lymphoma	Genentech, Biogen Idec	Rituxan
Envelope protein of the hepatitis B virus	Vaccination, Hepatitis B	SmithKline Beecham	Engerix-B

Bacterial Expression System

Among bacteria, *Escherichia coli* remains the most popular host of choice. Other bacterial systems in use are *Lactococcus lactis* and *Bacillus* sp. *E. coli* is non-pathogenic and can be grown rapidly in defined but non-expensive media. The genetics of *E. coli* has been extensively studied compared to other microorganisms so that transcription and translation processes are better understood. Additionally, *E. coli* cells can be easily broken and harvested to make protein isolation easier. Use of *E. coli* expression system is discussed in detail in the latter sections.

Table 2 Protein expression systems

Expression system	Host organism	Advantages	Potential challenges	Most common applications
Bacteria	• *E. coli* • *Lactobacillus lactis* • *Bacillus* sp.	• Simple and easy to grow • Low cost • Scalable • A number of cloning vectors available mammalian proteins	• Solubility of protein • Post-translational modifications minimum • Biological activity or immunogenicity of proteins may differ • High endotoxin in gram negative bacteria	• Antibody generation • Functional assays • Protein standards • Structural analysis • Preparation of isotope-labelled proteins
Yeast	• *Saccharomyces cerevisiae* • *Pichia pastoris* • *Pichia methanolica*	• Eukaryotic protein processing • Scalable to fermentation (grams per litre) • Requires simple inexpensive media • Safe	• Fermentation required for very high yields • Growth conditions may require optimization	• Proteins with desired glycosylation • Antibody generation • Vaccination • Protein interactions • Secreted proteins
Insect	• Baculovirus infection • Insect cell line	• Post-translational modifications similar to mammalian systems • Greater yield than mammalian systems • Safe	• More demanding culture conditions • Product not always fully functional	• Proteins with desired glycosylation • Functional assays • Structural analysis • Antibody generation • Secreted proteins
Mammalian	• Chinese hamster ovary cell line • Adenovirus, SV40 virus infection	• Correct post-translational modifications • Biological activity like native human protein	• Cells are difficult to grow • Expensive culture conditions • Low yield • Cells may be unstable following genetic manipulation	• Functional assays • Protein interactions • Antibody generation • Post-translational modifications study
Cell free	• Rabit reticulocyte lysate (red blood cells) • Wheat gram extract • *E. coli* extract	• Rapid expression • Open system—easy addition of factors for modulation of solubility or functionality • Simple • Scalable	• High expression yields (~3 mg)	• Rapid expression screening • Toxic proteins • Incorporation of unnatural labels or amino acids • Functional assays • Protein interactions

Yeast Expression System

The eukaryotic organism yeast can be cultured in defined and inexpensive medium and can be grown to high cell densities. These are useful for the production of eukaryotic proteins. In addition, yeast cells are adapted to fermentation so that large quantities of proteins can be produced by using yeasts. Three main species of yeast are in use: *Saccaharomyces cerevisiae*, *Pichia pastoris*, and *Pichia methanolica*.

Insect Cell Expression System

Insect cells, being higher eukaryotes compared to yeast, can carry out many processing events that occur in mammalian systems. For example, these insect cells can perform complex post-translational modifications, which are not possible in yeast, and also possess the machinery for appropriate protein folding, which yields soluble mammalian proteins. The vector system used for protein expression employs baculovirus. The most widely used example is the naturally infecting *Autographa californica*, a moth family insect, also called *Autographa californica* nucleopolyhedrovirus. The baculovirus is propagated in insect cell lines (SF 9 or SF 21) derived from *Spodoptera frugiperda*, belonging to the moth family.

Mammalian Expression System

Mammalian cells are often used for the expression of eukaryotic human proteins in order to get a product functionally closer to the original one. It is expected that post-translational modification and folding will be more efficient in mammalian cells compared to bacteria, insect or yeast host. For large-scale production, mammalian cells are less attractive because of relative difficulty in growing cells on a large scale and the high cost involved. Several viral-based expression vector systems are used such as the virus vectors from adenovirus, coronavirus, *Herpes simplex* virus, simian virus 40, and vaccinia virus. The baculovirus vector system can also be used in mammalian cells.

E. COLI PROTEIN EXPRESSION SYSTEM

The characteristics that make *E. coli* ideally suited as an expression system for many kinds of proteins are its rapid doubling time (approximately 30 min) in simple defined (and inexpensive) media. Availability of an extensive knowledge of its promoter and terminator sequences makes it suitable for expression of many proteins of both prokaryotic and eukaryotic origin. With careful manipulation, in some cases, strains producing 30% of their total protein as the expressed gene product can often be obtained (Box 1).

However, expression in *E. coli* does have disadvantages. First, being a prokaryotic organism, *E. coli* cells are unable to process introns, but the use of cDNA to produce an expression vector overcomes this problem. Second, *E. coli* cells do not possess the extensive post-translational machinery found in eukaryotic cells that can glycolylate, methylate, phosphorylate, or alter the initially produced protein in other ways, such as through extensive disulphide bond formation. Third, proteins expressed in large amounts in *E. coli* often precipitate into insoluble aggregates called "inclusion bodies", from which they can only be recovered in an active form by solubilization in denaturing agents, followed by careful renaturation. Fourth, it is relatively difficult to arrange the secretion of large amounts of expressed proteins from *E. coli*, although it has often been possible to secrete small amounts into the periplasmic space and to recover them by osmotic shock.

GENERAL CHARACTERISTICS OF THE EXPRESSION VECTOR USED IN *E. COLI*

The basic approach for the expression of all foreign proteins in *E. coli* begins with the insertion of the gene coding for the protein product into an expression vector, usually a plasmid. Much work has gone into the design of vectors for maximizing protein production. The architecture of a typical expression vector is shown in Figure 1.

An expression vector generally contains the following elements.

❖ A strong inducible promoter (for example, lac, trp, or tac), which can produce large amounts of mRNA from the cloned gene upon induction

❖ A transcriptional terminator for correct termination of mRNA

❖ An appropriately positioned ribosome-binding site and initiator ATG for efficient translation

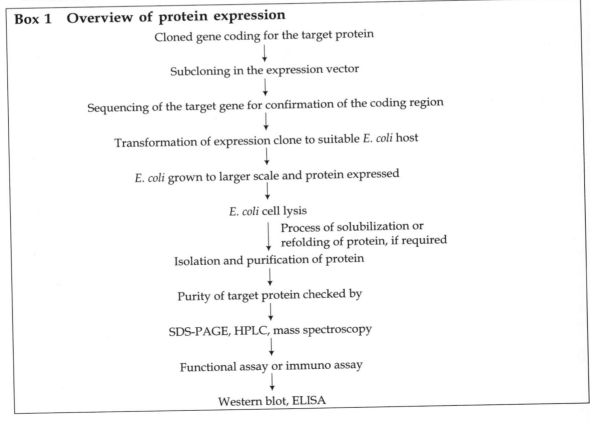

Box 1   Overview of protein expression

Cloned gene coding for the target protein
↓
Subcloning in the expression vector
↓
Sequencing of the target gene for confirmation of the coding region
↓
Transformation of expression clone to suitable *E. coli* host
↓
E. coli grown to larger scale and protein expressed
↓
E. coli cell lysis
↓ Process of solubilization or refolding of protein, if required
Isolation and purification of protein
↓
Purity of target protein checked by
↓
SDS-PAGE, HPLC, mass spectroscopy
↓
Functional assay or immuno assay
↓
Western blot, ELISA

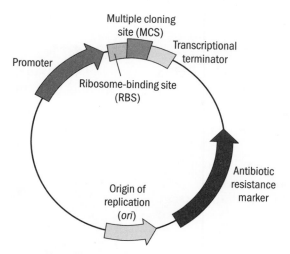

Figure 1   A typical expression vector

❖ A polylinker (multiple cloning site or MCS) located between the promoter and the terminator to ensure the insertion of the gene in correct orientation within the vector.

Once constructed, the expression vector containing the gene to be expressed is introduced into an appropriate *E. coli* strain.

Many proteins are toxic to the host bacteria. For example, the expression of poliovirus 3AB gene product is highly toxic to *E. coli* cells as it creates some drastic changes in the membrane permeability of the bacteria. Therefore, to maximize protein expression, it is vital that an inducible expression system be established, so that large quantities of the host cells can be grown before the expression of the target protein is initiated. Protein production can then be activated rapidly and the cells can be harvested soon afterwards prior to the potentially toxic effects of the expressed protein.

OPTIMIZATION OF EXPRESSION LEVELS IN *E. COLI* EXPRESSION SYSTEM

The expression of a foreign gene from eukaryotic source in a prokaryotic organism *E. coli* is not straight forward. The expression of a gene involves transcription and translation, which are in turn regulated by signals that might be different for prokaryotes and eukaryotes. Other factors like base composition and codon usage also come into play. The optimization of gene expression can be done considering the effect. Also the optimization of the factors discussed can be subsequently attained.

Promoter Optimization

Promoter is the region located at the 5′-flanking region of a gene, where RNA polymerase binds and initiates RNA synthesis. Promoter determines as to which DNA regions need to be expressed and how much the expression would be. A strong promoter produces higher expression of a gene while a weak promoter causes lower expression. This careful regulation of gene expression by promoter is controlled by the sequences of the promoter region and the sigma factor of RNA polymerase that determines the specificity. It is thus expected that a gene from one host may not be expressed in another under normal conditions.

Eukaryotic organisms possess promoters that are significantly different from bacterial ones and the eukaryotic system has the mechanism to remove the introns present in the eukaryotic genes, which are lacking in the prokaryotic system. These problems can be solved by cloning cDNA, instead of cloning genomic DNA. In such a case, the upstream region is not present in the cloning fragment. An external promoter is provided in the expression vector to carry out successful expression.

PROMOTER SEQUENCES

Many different promoter sequences have been used to elicit inducible protein production in *E. coli*. Some of these are discussed as follows.

The lac Promoter

The *E. coli* lac promoter has been used because it provides a mechanism for inducible gene expression. In many cases, the eukaryotic proteins produced in bacteria may be toxic, and even if they are not toxic, excess production may interfere with the bacterial growth. In either case, a system is desired, which can keep the clone gene repressed and turn it on when desired. The lac promoter provides such a system. This is possible by exploiting the fact that the lac genes are expressed maximally when *E. coli* is grown on lactose or induced by a lactose analog isopropyl-β-D-1-thiogalactopyranoside (IPTG). If the lac promoter sequences are placed upstream of a cloned gene, the latter will be turned on in a lactose-dependent fashion. The lac promoter can thus be kept off and stimulated by IPTG after sufficient growth of the bacterial cell.

The lac promoter has the following advantages.

❖ It is inducible in nature.
❖ It can be readily activated.

But lac promoter alone is rarely used because of the following drawbacks.

❖ The lac promoter is fairly weak and, therefore, cannot drive very high levels of protein production.
❖ Due to the absence of *lacI* gene, the repression is incomplete and the lac genes are transcribed to some extent before induction.

The latter problem can be partially overcome by expressing mutant versions of the *lacI* gene, which have increased DNA-binding (and consequently repressing) ability—for example, the *lacI^q* allele results in the overproduction of *lacI* and, consequently, results in a reduced level of transcription in the absence of inducer.

The Tac Promoter

The tac promoter was designed to replace the weak lac promoter, causing less target protein product. The analysis of the promoter regions of a large number of genes from *E. coli* revealed that there exist two consensus domains at –35 and –10 regions (meaning that the regions start from 35 and 10 bases upstream of the transcription start site). RNA polymerase binds at these two sites during the initiation of transcription of the gene under consideration. The lac promoter sequence deviates from the consensus at –35 region (Figure 2), and, therefore, it is presumed that RNA polymerase cannot bind strongly to this region. Alternatively, the –35 region of another promoter trp was found to have the consensus sequence conserved. The trp promoter is, therefore, a strong promoter, which is known to control the expression of the genes responsible for tryptophan biosynthesis and response to tryptophan limitation. The tac promoter is a hybrid promoter constructed by the fusion of –35 region of the *E. coli* trp promoter with –10 region of lac promoter. It, therefore, possesses the properties of both the promoters.

The tac promoter is able to induce the expression of target genes such that the encoded polypeptide can accumulate at a level of 20%–30% of the total cell protein. Like lac promoter, expression of the tac promoter can be repressed by *lacI* protein. A mutated form of *lacI* gene called *lacI^q* when present represses the tac promoter. Since addition of IPTG can inactivate *lacI*, the amount of expression from tac promoter is proportional to IPTG concentration. On the other hand, high expression from tac promoter is obtained at high concentration of IPTG and consequently the expression of the gene under tac promoter.

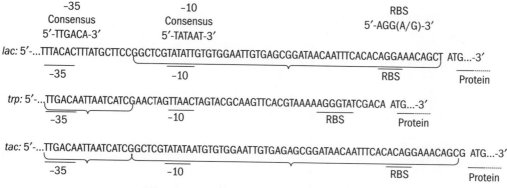

Figure 2 Protein expression promoter

The disadvantages of tac promoter are as follows.

❖ It requires the presence of *lacI^q* on the same vector.

❖ Excessive overexpression may inhibit cell growth or result in the formation of insoluble inclusion bodies or protein aggregates. If not fully repressed, the system can be leaky leading to some expression of the repressed gene leakiness.

The Phage Promoter

Another strategy that has been widely used for protein overexpression is that the gene under consideration is placed under the control of a phage-regulated promoter. One of them is λP_L promoter. This promoter is responsible for transcribing the genes present on the left side of the λ genome, namely, N, cIII, and so on; the cI gene product represses the promoter. This promoter can be activated by a temperature-sensitive mutant of cI (mutation is effective at a particular growth temperature) called cI857. When the growth temperature is changed to 30°C, cI repressor is functional and the gene expression under the control of λP_L promoter is turned off. But at 42°C, the cI repressor is inactivated and the expression of the gene is turned on. P_L vectors are marketed by Invitrogen, Life technologies, USA.

Alternatively, phage promoter recognizing phage-encoded RNA polymerase can be used. The promoter sequence is different from *E. coli* promoter sequences. Expression vectors use any of the three phage-specific RNA polymerase promoter systems, namely, T7, SP6, and T3. These systems can generate very high level of transcription of the downstream gene. The high level transcription achieved by these vector systems may sometimes be toxic to the host. To avoid such problems, an inducible system needs to be introduced. The phage RNA polymerase may be cloned downstream of a regulated promoter, or the polymerase may be introduced into the cell with the help of a defective phage as in the T7 expression system (Box 2) used to construct pET vectors.

OPTIMIZATION OF OTHER FACTORS INFLUENCING PROTEIN EXPRESSION

Promoter strength is not necessarily the determining factor for the accumulation of the target protein within the cell. It has been found that many proteins cannot be produced in *E. coli* cells even if excellent promoters have been used. So there are some additional factors that need to be considered for the successful expression of proteins. Some of these are discussed as follows.

Conditions of Induction

In spite of having sufficiently strong transcriptional signal, the level of expression of the desired protein may be low. This can be modulated by modulating the time of induction or changing the concentration of the inducer.

Box 2   T7-based pET expression system

The T7-based pET expression system (Plate 1) is marketed by Novagen, Madison. WI is most widely used for recombinant protein preparation. The pET vector is a bacterial plasmid designed for production of large quantity of protein products on activation. This plasmid contains the following elements: *lacI* gene coding for lac repressor protein, an ampicillin resistance marker, T7 transcription promoter, a polylinker region, lac operator which can block transcription, f1 origin of replication which enables the production of single-stranded vector when required, and the conventional ColE1 origin of replication.

The target genes are positioned downstream of the bacteriophage T7 RNA polymerase promoter. Propagation of this vector in wild-type *E. coli* cells will not result in the expression of the target gene since the T7 RNA polymerase is absent. T7 RNA polymerase is different from *E. coli* RNA polymerase in the sense that the latter does not recognize T7 promoter sequences as start sites for transcription. To elicit target gene expression, an *E. coli* cell is constructed in such a way that a copy of the gene for T7 RNA polymerase (T7 gene 1) is integrated into its genome under the control of the lac promoter. *E. coli* BL (DE3) represents such a host. Additionally, the promoters for both the target gene and T7 gene 1 also contain the *lacO* operator sequence and are, therefore, inhibited by the lac repressor (*lacI*). Isopropyl-β-D-1-thiogalactopyranoside (IPTG) induction allows the transcription of the T7 RNA polymerase gene whose protein product subsequently activates the expression of the target gene. To control the leaky production of T7 RNA polymerase (thereby ensuring that target gene expression is minimized), *E. coli* cells can be co-transformed with an additional plasmid containing compatible replication origin to the expression vector. This additional plasmid-producing T7 lysozyme, when introduced in the *E. coli* cell, inactivates any T7 RNA polymerase that may be produced in the absence of induction. The production of this inhibitor will inactivate the small levels of polymerase produced in the absence of induction, but it will be swamped and thereby rendered ineffective by the larger amounts of polymerase produced during induction.

Level of Expression Vector

To achieve high expression of the target gene, high copy number plasmids (vectors) are chosen. Usually, vectors present in 15–20 copies (for example, ColE1 or pMB1 derived vectors) or a few hundred copies (pUC series of vectors) are chosen. When co-overexpression of additional genes is required, ColE1 derivatives are combined with compatible plasmids (for example, p15A replicon) and maintained at 10–12 copies per cell. In some cases, a very high expression vector copy number did not result in increased protein production and in others, increased vector levels actually reduced the levels of protein production. Most commercially available expression vectors today contain the replication origin of either pBR322 or pUC, and altering copy number is not commonly used to modulate protein production.

Transcriptional Termination

Efficient transcription termination is an essential component for achieving high levels of gene expression. A transcription terminator is placed downstream of the sequence encoding the target gene. Transcription terminators serve to enhance plasmid stability by preventing transcription through the origin of replication and from irrelevant promoters located in the plasmid. Further, terminators stabilize the mRNA by forming a stem loop at the 3'-end. This can lead to substantial increase in the levels of accumulated protein. The two tandem transcriptional terminators (T1 and T2) from the rrnB rRNA operon of *E. coli* are often present in expression vectors but other terminators also work well.

Codon Usage

The degeneracy of the genetic code means that there are several codons that code for the same amino acid. All these codons are not equivalent. The cell will use some codons more readily than others. The genes of both prokaryotes and eukaryotes show a non-random usage of alternative codons. Genes containing favourable codons will be translated more efficiently than those containing infrequently used codons. In general, the frequency of the use of alternative codons is correlated with the relative abundance of their cognate tRNA molecules.

Differences in codon usage between prokaryotes and eukaryotes can have a significant impact on the heterologous protein expression. For example, the arginine codons AGG and AGA are infrequently used in *E. coli* whereas they are common in *S. cerevisiae* and eukaryotes. The presence of such codons in a cloned gene could be a limiting factor in the bacterial production of several mammalian proteins. Moreover, it may affect mRNA and plasmid stability and, in some cases, may inhibit protein synthesis and cell growth. This problem can be addressed by using site-directed mutagenesis to replace the rare arginine codons by *E. coli* preferred CGC codon (without changing the polypeptide sequence). Alternatively, the co-expression of the gene coding for tRNAArg(AGG/AGA) (dnaY) will produce the cognate tRNA and can result in high-level production of the target protein.

The above two approaches are routinely used to take into account the differential codon usage. Systems have been established for the expression of other tRNA molecules that occur frequently in mammalian coding sequence but are used rarely in *E. coli*. One such system uses a bacterial strain (called RosettaTM) that expresses the tRNAs for AGG, AGA, AUA, CUA, CCC, and GGA on a plasmid compatible with the expression vector. Site-directed mutagenesis is used to alter the gene that is to be expressed such that the codons it contains are more frequently used by other highly expressed genes in *E. coli*. That is, the DNA sequence of the gene is altered to allow more favourable codons to be used but the encoded polypeptide remains unchanged.

GC Content

The base composition of organisms vary widely so that the GC content of the cloned gene may be substantially different from that of the host in which it has been cloned. The difference in the base composition might have an effect on the transcription, codon usage, structure and stability of the mRNA, and also on the stability of the vector itself. Specifically, high GC content at the 5′-end of the cloned gene produces secondary structure in the mRNA, which might cause lower level of expression and can be corrected by replacing with A or T residues at the 5′-end without changing the amino acid coded.

Addition of Fusion Partners

The protein to be expressed may be fused with an amino acid tag at the N-terminal, which is often useful for protein purification. Another strategy employs the fusion of the N-terminal region of the protein with the C-terminal of another highly expressed protein, resulting in a high-level expression of both partners.

Use of Protease-deficient Host Strains

The host strains are often mutated for protease production so that there is a reduction in the proteolytic degradation. As a result, there will be enhanced accumulation of the heterologous protein. The strain BL21, widely used in *E. coli* expression system, has mutation in two protease genes *lon* and *ompT*.

STRATEGIES FOR IMPROVING PROTEIN SOLUBILITY

A problem commonly encountered in the expression of heterologous proteins is that of protein solubility. If a protein is not properly folded, the folding intermediates often aggregate and become insoluble. Insoluble proteins are packed to form a complex structure called inclusion bodies. Several strategies are used to improve the solubility of the heterologous protein.

Reducing the Rate of Protein Synthesis

Some proteins become insoluble because they are synthesized so rapidly that they cannot be correctly folded. Reducing the rate of protein synthesis often improves the solubility problem. The synthesis rate can be reduced by the following strategies.

❖ The bacteria are grown at a lower temperature, which will decrease the rate of protein synthesis in general, and effectively soluble proteins can be obtained.

❖ The heterologous DNA may be cloned in a low copy number vector.

❖ A weaker promoter can be used so that the transcription rate is reduced.

❖ The concentration of the inducer can be decreased, resulting in lower induction and, hence, lower synthesis of protein.

Optimizing the Growth Medium

Some proteins become insoluble because of the absence of necessary cofactors or prosthetic groups like metals, minerals, vitamins, and so on in the growth medium. These are required for correct folding and, hence, absence of these affects the folding, making the protein insoluble. So the growth media can be optimized in the following ways.

❖ Addition of trace metals, minerals, vitamins, which serve as cofactors or prosthetic groups essential for the proper folding of the heterologous protein

❖ Adjustment of pH by the use of appropriate buffer composition

❖ Addition of sugars for repression (for example, addition of 1% glucose is effective in repressing the induction of the lac promoter by lactose), carbon sources as well as osmoprotectants (these are the molecules created within bacterial cells when the outside osmolarity increases, and these molecules have a role in the stabilization of protein structures)

Co-expression of Molecular Chaperones and Foldases

Protein folding in vivo requires molecular chaperones or foldases or both. Molecular chaperones bind to misfolded or unfolded proteins and refold them back to their quaternary structure. Examples of molecular chaperones in *E. coli* include GroEL-GroES, DnaK-DnaJ-GrpE, ClpB. Foldases are enzymes that catalyse the formation and isomerization of the disulphide bonds in protein and have a role in accelerating protein folding. Examples are peptidyl prolyl cis isomerase, peptidyl prolyl trans isomerase (PPI), DsbA-DsbB-DsbC—the enzymes of thio-oxidoreductase family—protein disulphide isomerase (PDI), and so on. If one or more of these proteins are co-expressed with the target protein, higher amounts of soluble proteins are obtained. However, it is necessary to optimize the levels of co-expression, and this has to be done for each protein to be expressed.

Expressing the Protein in the Periplasmic Region

Many of the bacterial proteins are folded in the periplasmic region under normal conditions. Thus the heterologous proteins, if expressed in the periplasmic region, are expected to have some folding advantages. The distinct advantages of the periplasmic compartment are as follows.

❖ The formation of disulphide bonds is easier in the oxidizing environment of the periplasm compared to that in the cytoplasmic region.

❖ Foldases are present in the periplasmic space, which can be utilized for folding.

❖ Accumulation of toxins is possible (the toxins of pathogenic bacteria often reside in the periplasmic space).

❖ Less proteases are present.

For secretion to the periplasmic space, a small peptide sequence called signal peptide is needed at the N-terminal of the protein. The signal sequences from the proteins PelB or OmpT are utilized for this purpose. Although protein expression at the periplasm promotes stability and reduced toxicity, the yield becomes fairly low compared to the cytoplasmic expression.

Use of Specific Host Strain

Specific strains have been constructed to increase the solubility of proteins with disulphide bonds in the cytoplasm. One of these is AD494 derived from *E. coli* having a mutation in *trxB* gene coding for thioredoxin reductase. The mutation is responsible for reduced thioredoxin reductase activity and enhances cytoplasmic disulphide bond formation. Another strain, a double mutant of *trxB* and glutathione reductase gene (*gor*), is also used. The above two strains are marketed by Novagen.

Adding a Fusion Partner

If the N-terminus of protein to be expressed is fused to the C-terminus of a soluble protein, the solubility of the target protein improves. Some proteins are expressed in soluble form as a complex comprising more than one subunit, so that these subunits are their natural partners. To increase the solubility of a particular subunit, it is often co-expressed with its natural partner.

Expression of the Fragment of a Protein

Large proteins are difficult to express in *E. coli* and also difficult to make them soluble. So smaller fragments are often expressed. The solubility of a protein can be improved by selecting the hydrophilic domain for the expression.

In Vitro Denaturation and Refolding of the Protein

Sometimes it is not possible to recover the target protein in soluble form despite all efforts. In vitro methods are then attempted. This can be done through the following steps.

❖ Inclusion bodies are isolated.

❖ Denaturing agents are added (guanidine hydrochloride or urea) under reducing condition (in presence of dithiothreitol, or DTT).

❖ Denaturing agent is removed by dialysis (diffusion through a specific membrane) when refolding occurs.

IMPROVING PROTEIN STABILITY

The proteins that are folded properly are, in general, stable. The structural features responsible for the stability of a protein are not precisely known. There are, however, some general criteria that can be considered for modulation of stability. These are as follows.

❖ *N-end rule*   The half-life of a protein in *E. coli* decreases if the N-terminal end of the protein contains the amino acid residues Arg, Lys, Leu, Phe, Tyr, or Trp.

❖ *PEST hypothesis*   If eukaryotic proteins possess regions enriched in amino acid residues Pro, Glu, Ser, or Thr, they are destabilized.

❖ Protease-deficient host strains can also be used for increasing the stability (decreasing proteolysis).

❖ Periplasmic expression increases stability as proteolysis is less in the periplasm.

❖ Reduction in growth temperature is used for increasing stability. At lower growth temperature, the protein expression will be slower but proteolytic degradation will also be slower; a balance between the two may result in higher protein production.

REDUCING PROTEIN TOXICITY

The expression of a recombinant protein is foreign to the host in which it is expressed. The overproduction of such a protein is often toxic to the host cell because under normal physiological condition, this foreign protein will perform certain functions that may not be needed by the host. It may cause low growth rate of host cells or may often be detrimental to the host. Protein toxicity also affects the cloning and expression of a target protein. Expression problems manifest as no expression, or low yield of protein, or formation of defective protein, or inconsistent expression of the target protein.

Depending on the toxic manifestations, strategies can be devised to minimize the toxic effects. The salient points are discussed subsequently.

Incomplete Repression of Protein Expression before Induction

As some promoters (for example, lac promoter) are leaky and are not tightly regulated, some amount of protein is expressed before induction. This may lead to plasmid instability and loss of plasmids. To have a more tightly regulated expression, the following strategies could be undertaken.

❖ Use a tightly regulated promoter like the arabinose promoter.

❖ Use vectors with lower copy number.

❖ Create constitutive expression of a repressor protein.

❖ Use glucose in the media to repress the induction of lac promoter, which requires lactose for induction.

Toxicity Following Induction

To decrease the effect of toxicity for those proteins that either inhibit the growth of the host cell upon induction or kill the host, the following approaches may be used for protein expression.

❖ *Periplasmic expression of target protein*   The protein may be secreted to the periplasm, which can hold toxic proteins, and in this way, cytoplasmic expression is avoided.

❖ *Expression of proteins in inclusion bodies*   To direct such expression is difficult, but if the protein exists in an aggregate, it may not be toxic to the host.

Expression of Non-toxic Domains

In case of toxic proteins, domains exist; this may not be toxic like the whole protein and can be expressed.

Use of Special Host Strains

Some host strains are constructed that can handle the toxic proteins better. Examples are C41(DE3) or C43(DE3), which are better for membrane protein expression compared to their parent strain BL21(DE3).

IN VITRO EXPRESSION USING *E. COLI* EXTRACTS

In vitro protein expression systems involve the production of recombinant proteins in the solution by the use of in vitro transcription followed by translation with the translation machinery of *E. coli* extract. In this method, protein synthesis takes place in cell lysates rather than inside the organism *E. coli*. This is called cell-free method. The method is faster compared to in vivo methods. The cell-free protein expression systems are marketed by several companies such as Roche, Invitrogen, Novagen, and Qiagen and are widely used for protein production.

The cell-free protein synthesis requires mainly the template DNA, which codes for the target protein and a solution containing the required components for transcription and translation. The cell extracts provide the molecules required in the reactions such as RNA polymerase for transcription, ribosomes, tRNA, amino acids for translation, and also supply the enzyme cofactors, energy sources, and the components required for proper folding of the target protein. Cell-free expression system can be either linked or coupled. In the linked system, the DNA template is transcribed at first and then a part of the transcribed product is added to the translation mixture. On the other hand, in a coupled reaction, both the transcription and translation reactions are carried out in the same tube where all the requisite reaction components are present.

The advantages of cell-free systems are as follows.

❖ The reaction is faster compared to in vivo protein expression.

❖ Expression of toxic proteins is not a problem.

❖ Isotope labelling of proteins can be done.

❖ Membrane protein production can be done with some modifications.

However, cell-free systems from *E. coli* are deficient in post-translational modifications and have lower yield of proteins in some cases.

CO-EXPRESSION

Co-expression of target proteins with another suitable protein can be used for proper folding or solubilization of the target protein. For example, a protein having three subunits is soluble but each subunit is not. This could be due to the presence of exposed hydrophobic regions in each subunit, which become hidden during complex formation. To express one such subunit, one may choose to co-express with others so that the protein is obtained in a soluble form. Another example is to co-express foldases or chaperones along with the target protein collected in a soluble form.

Methods of Co-expression

Co-expression of two or more proteins can be done as follows.

❖ *Co-expression from two different vectors* Vectors expressing two proteins are located inside the same host. Here two plasmids must coexist inside the host for which the vectors must contain different antibiotic resistance markers and origin of replication should also be different.

However, since copy number of two vectors may differ, there would be differences in the expression level of the co-expressing proteins. The expression level can be optimized by choice of suitable promoters to express the genes in two different vectors.

❖ *Co-expression from one vector* In this case, two genes coding for the target proteins are cloned in the same vector and are placed under the control of the same promoter (Figure 3).

Again, if the genes are cloned under the same promoter, the order in which the genes are present will modulate the expression level of the proteins. To optimize such a situation, two constructs may be prepared varying the order of the two genes.

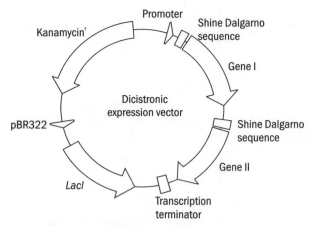

Figure 3   Dicistronic expression vector

PROTEIN PURIFICATION

The protein expression systems efficiently convert a gene into a protein so that the protein is highly expressed inside a suitable host. The next step will be to separate the target protein from thousands of protein present in the host cell. The separation and identification of the foreign protein can be done by high-resolution two-dimensional sodium dodecyl sulphate–polyacrylamide gel electrophoresis (SDS-PAGE). In this technique, the proteins are separated by charge in the first dimension and by size in the second dimension. A large number of spots are usually obtained, and the separation may not be easy. However, in two-dimensional electrophoresis, proteins are denatured and may not be suitable for functional studies.

Traditional purification methods can be used such as gel filtration (separation on the basis of size), ion-exchange chromatography (separation based on charge) or hydrophobic interaction chromatography (separation based on hydrophobicity of proteins). However, if the properties of the foreign protein are similar to that of the host protein, it will be difficult to identify and separate them by any of the above methods.

To overcome the above problem, it is required that the target protein be modified in such a way that it acquires some unique property (by which it can be identified), but at the same time, the function is not impaired. The concept of protein purification tags solves the problem. Protein purification tags are small peptides possessing specific high-affinity binding properties, by virtue of which they are allowed to bind to a solid support (Figure 4).

Ideally, the tag should be such that the recombinant protein will bind to the column with high affinity and specificity, while other proteins are washed out. But at the same time, it is desirable that the tagged protein is eluted in a native state.

The His tag provides one such example with which a foreign protein can be purified easily. In this procedure, a sequence coding for a six-consecutive histidine residue is added to one end of the insert that codes for the target protein. The target protein, when expressed, will thus contain 6 His residues at its N- and C-terminal ends. The histidines are capable of binding non-covalently with high affinity to Ni^{2+} ions. His-tag proteins are then passed through a Ni^{2+} resin when the tagged protein is retained by the resin, but the other proteins are washed through. The commonly used Ni^{2+} resin is nitriloacetic acid (NTA) covalently attached to the nickel.

The advantages of using His tags are as follows.

❖ The small size of the tag does not affect immunogenicity of the target protein.

❖ His tag may be placed at N- or C-terminus.

❖ His tag-Ni^{2+} interaction does not depend on the tertiary structure of the tag; so the protein can be purified by using denaturing condition, if required. Generally, denaturing conditions are required for the purification of otherwise insoluble proteins.

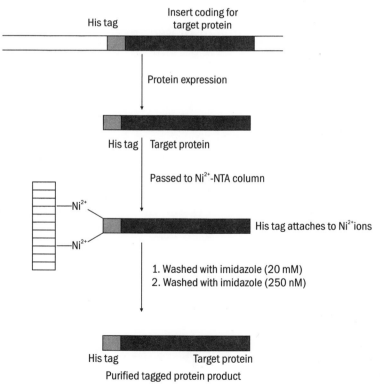

Figure 4   Protein purification by tagging

Table 3 Commonly used fusion tags for protein expression and purification

Tag	Size	Tag placement	Use
His tag	6–8 amino acids	N-, C-terminal, Internal	Purification by Ni-NTA column
FLAG tag	8 amino acids	N-, C-terminal	Purification by immobilized specific antibody
HA tag	9 amino acids	N-, C-terminal	Purification by immobilized specific antibody
c-Myc tag	11 amino acids	N-, C-terminal	Purification by immobilized specific antibody
Strep tag II	8 amino acids	N-, C-terminal	Purification by Strep-Tactin ligand immobilized on a Sepharose base
SBP tag	38 amino acids	C-terminal	Purification with immobilized streptavidin
Chitin binding domain	51 amino acids	N-, C-terminal	Purification with chitin resin
Calmodulin binding domain	26 amino acids	N-, C-terminal	Purification with calmodulin resin
Cellulose binding domain	27–129 amino acids	N-, C-terminal	Purification with cellulose column
GST tag	220 amino acids	N-terminal	Solubility enhancement of protein and purification by immobilized glutathione agarose column or beads
MBP tag	392 amino acids	N-, C-terminal	Solubility enhancement of protein and purification by immobilized starch
NusA	495 amino acids	N-terminal	Solubility enhancement of protein
Thioredoxin (Trx)	109 amino acids	N-, C-terminal	Solubility enhancement of protein
SUMO	100 amino acids	N-terminal	Solubility enhancement of protein

Some commonly used protein purification tags are listed in Table 3.

SUMMARY

- Protein expression system consists of suitable vector and host and is used for the expression of proteins in substantial amounts by cloning the desired gene.

- The protein expression in bacterial, yeast, insect, or mammalian systems depends on the host and can also be cell free.

- *E. coli* expression system uses vectors containing strong inducible promoters like lac, tac, or trp, and the gene encoding the protein is placed under its control for overproduction. The expression level of the protein is optimized by choosing appropriate conditions for promoter induction. The codon usage and GC content of the cloned gene are also optimized for heterologous protein expression.

- Appropriate protein folding and protein solubility can be ensured by the reduction in the rate of protein synthesis, use of appropriate growth medium, or addition of chaperones or using protease-deficient host cells. Co-expression of target proteins with another suitable protein can also be used for proper folding or solubilization of the target protein.

- Protein toxicity can be reduced by expressing proteins in the periplasmic space or inside inclusion bodies.
- In vitro protein expression systems involve the production of recombinant proteins in solution by using in vitro transcription followed by translation with the translation machinery of *E. coli* extract.
- The highly expressed target protein is purified by the use of protein purification tags. These are small peptides possessing specific high-affinity binding properties by virtue of which they are allowed to bind to a solid support. The recombinant protein, being tagged, binds to the column with high affinity and specificity, while other proteins are washed out. The tagged protein is then eluted in a native state.

REVISION QUESTIONS

1. Expression vectors are widely used in recombinant DNA technology. What is an expression vector? What are the essential features of an expression vector? What special properties an expression vector must have over a cloning vector?
2. What is the advantage of expressing a protein in mammalian cells versus bacteria?
3. Describe the reasons why it may be useful to clone a gene or a cDNA into an expression vector.
4. What are the potential advantages and disadvantages of using an expression vector with a strong constitutive promoter?
5. Why is an inducible promoter used in an expression vector?
6. What is cDNA? Why is it important to use cDNA when trying to express a eukaryotic protein in a prokaryote? How can one make cDNA?
7. Explain how an expression vector gets around the problem of differing promoters in prokaryotes and eukaryotes.
8. How can one modify the foreign protein to facilitate its purification?

8

Site-directed Mutagenesis

OBJECTIVES

After reading this chapter, the student will be able to:
- Describe site-directed mutagenesis using single-stranded DNA
- Explain primer extension mutagenesis or oligonucleotide-directed mutagenesis
- Discuss oligonucleotide-directed mutagenesis with phosphothioate strand selection or Eckstein method
- Explain oligonucleotide-directed mutagenesis with dut-ung- (or Kunkel) strand selection
- Describe cassette mutagenesis
- Analyse PCR-based mutagenesis
- Specify the considerations for design of mutagenic oligonucleotides

INTRODUCTION

Once the sequence of a gene is known, the amino acid sequence of the encoded protein can be determined. Accordingly, any mutation in the gene (alteration of DNA sequence) may result in changes in the amino acid sequence. To elucidate the function of a protein, analysis of mutations is very useful. Mutations may reduce the activity of the protein or impart abnormal properties to it. Once the function of the protein is known, one would like to find out regions of the protein that are important for its function. In case of enzymes, one would like to know the active site. Also one may like to know the contribution of individual amino acids or a group of amino acids for the structure and function of a protein. To get such insights, we need to alter one or a few amino acids of the protein and this can be done by mutating the gene.

Mutations may occur naturally due to erroneous DNA replication, but this error rate is low (~ one base alteration per 106 to 108 base extension during replication) and wide variation occurs among genes as well as among organisms. Initial studies relied on natural mutants to isolate genes and elucidated specific functions for the encoded protein. Before the advent of molecular biological techniques, mutation rates were increased by treatment of cells with specific physical or chemical agents. For example, when cells were treated with X-rays, UV rays, or with chemicals like nitrosoguanidine, ethyl methane sulphonate, bromouridine, 2-aminopurine, and so on, changes in the DNA sequence occurred at a faster rate. The mutations acquired in this way will be passed on to generations thereafter. The physical treatment (radiation) causes damage to the cellular DNA, and during repair, some mistakes generally occur resulting in the incorporation of improper bases. The chemicals used may serve as base analog or modify DNA chemically or can intercalate between bases. All these will have an effect on the activity of DNA polymerase, which will be prone to making mistakes. As a result, there will be base substitution (replacement of one base by another), insertion (addition of one or more bases), and deletion (removal of one or more bases).

The traditional method of mutagenesis suffers from the following drawbacks.

❖ The mutation can occur anywhere in the genome and cannot be controlled to be restricted to a specific gene.

❖ Isolation of a mutant phenotype does not guarantee mutation in the desired gene.

❖ Prior to the era of cloning and sequencing, it was not possible to determine the exact nature of the base change (whether it is a single base change or insertion or deletion of a fragment of DNA).

The other methods of mutagenesis, which gradually came up, were insertion mutagenesis that signify the insertion of one base or a sequence of bases in the genome resulting in the inactivation of function. Insertion mutations often occur naturally, such as those caused by transposons or viruses or specifically designed such as signature-tagged mutagenesis (STM).

Transposons are a class of mobile genetic elements (short DNA sequences) that can jump from one position within the genome and integrate to another position by a process called transposition. Transposons were discovered by Barbara McClintock in 1948 in maize while she was working in her Cold Spring Harbour Laboratory, USA and she was awarded the Nobel Prize in 1983 for this discovery. The transposition event does not rely on sequence specificity, and therefore there is no specific site for its insertion. A transposon usually contains genes involved in regulating the movement of transposon and a selectable marker, which is usually an antibiotic resistance marker. In addition, each transposon has direct or inverted DNA repeats located at its two ends. The structure of a typical transposon Tn5Tp is shown in Figure 1.

Whenever a transposon is inserted into a gene, the latter is functionally impaired. Thus transposition event may cause movement of host sequence from one place to

Figure 1 Schematic representation of a transposon Tn5Tp containing trimethoprim resistance derived from Tn5

another. This property of transposons has been utilized for generation of mutants in a technique known as STM. In this technique, the transposon carries a unique signature tag (unique short sequence that can later be identified through hybridization analysis), and mutations are generated by random insertion of transposon. An example of screening virulence genes by STM is illustrated in Box 1.

Viral insertion mutagenesis describes the phenomenon of integration of part or whole of viral genome sequences into the host through viral infection, resulting in the alteration of functions. In many cases, the viral genome is integrated into the host chromosome, and this is a fairly common cause of mutation. Insertion of viral genome often causes some deleterious effects; for example, when inserted in or near a gene controlling cell growth, it may result in oncogenic transformation in human. Hepatitis B virus (HBV) uses a similar mechanism in developing hepatocellular carcinoma in humans.

Box 1 Signature-tagged mutagenesis for the identification of virulence genes

The major breakthrough leading to the discovery of a group of genes encoding a specialized secretion system for the delivery of bacterial effector proteins directly into the host cells came in 1995 when signature-tagged mutagenesis (STM) was used for the identification of virulence genes in the bacterium *Salmonella typhimurium*. The strategy uses a signature tag wherein a central unique sequence (~40 kb) is flanked by two invariable sequences at both sides (Plate 1). Several tags are constructed, each with a unique sequence but having the same flanking sequences. The tag can be amplified with primers designed from the invariable sequences and then radiolabelled by γ^{32}P-ATP. A number of tags are then pooled and ligated to transposons. These transposons are then inserted into the bacteria, and the resulting mutants are assembled into a library. Mutant bacteria containing unique tags are then pooled and used to infect mice. This group of bacteria is termed input pool. The bacterial pool recovered from mice is further designated as output pool and represent the virulent variety. Genomic DNA was isolated from each of these, and the tags were amplified by polymerase chain reaction (PCR) and radiolabelled. These radiolabelled tags were used as probes in a colony hybridization experiment. The colony that hybridizes to the input but not to the output pool is regarded as non-virulent. Thus this is a negative selection method.

With the advent of molecular biological techniques, the in vitro mutagenesis methods were refined and improved. The most remarkable breakthrough came with the discovery of site-directed mutagenesis (SDM) for which the 1993 Nobel Prize in chemistry was awarded to Smith and Mullis for developing the polymerase chain reaction (PCR) technique. STM is the process by which a single base or a short stretch of bases can be altered in a cloned gene. In the SDM method, instead of generating mutations in organisms and then screening large number of progenies to get the desired mutant, it is feasible to change any base in a cloned DNA sequence. It can be regarded as a valuable tool in the manipulation of genes and structure function studies. In addition, this technique enables researchers to design and engineer any protein and impart novel properties or add values to these protein products, giving rise to a new discipline called protein engineering.

SITE-DIRECTED MUTAGENESIS USING SINGLE-STRANDED DNA

Since the discovery of the powerful technique of SDM, a number of procedures have been described in the literature. Some of the widely used procedures are as follows.

❖ Primer extension mutagenesis or oligonucleotide-directed mutagenesis
❖ Oligonucleotide-directed mutagenesis with phosphothioate strand selection, or Eckstein method
❖ Oligonucleotide-directed mutagenesis with dut-ung- (or Kunkel) strand selection
❖ Cassette mutagenesis
❖ PCR-based mutagenesis

Primer Extension Mutagenesis or Oligonucleotide-directed Mutagenesis

The primer extension mutagenesis method uses oligonucleotides in creating site-directed mutations and was devised by Michael Smith. This method of mutagenesis takes advantage of the properties of bacteriophage M13. M13 undergoes a switch during its life cycle when its single-stranded genome is converted to a double-stranded form. In other words, the single-stranded form of the genome serves as a template for the new synthesis of a second DNA strand. This procedure is also called oligonucleotide-directed mutagenesis. The steps involved are explained as follows (Figure 2).

Step 1 *Primer annealing to template DNA* The gene or the DNA fragment is inserted in an M13 phage vector. The single-stranded DNA is isolated from the recombinant M13 phage and used as a template for the binding of an oligonucleotide primer. The primer is designed in such a way that it differs in one base (depending on the mutated base that needs to be introduced). Since the single-stranded molecules have the ability to anneal to each other even if there are some mismatches, the primer anneals to the template and serves as the primer for DNA synthesis by DNA polymerase.

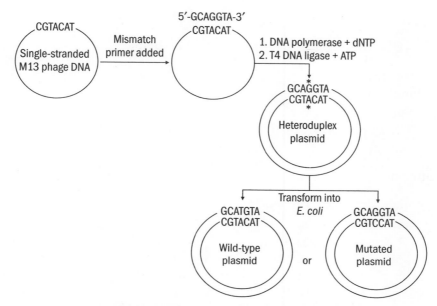

Figure 2   Primer extension mutagenesis

Step 2 *Primer extension by DNA polymerase and synthesis of second strand*   The hybrid is then treated with DNA polymerase in the presence of four deoxynucleotide triphosphates (dNTP) to synthesize a new M13 DNA strand complementary to the original at every base, except for those alterations introduced in the primer.

Step 3 *Sealing of nicks by ligase*   The sugar–phosphate backbone of the new DNA circle is then completed using DNA ligase. DNA ligase seals the nick and a completely closed, circular double-stranded DNA molecule is produced. One strand of this molecule contains the original sequence while the other strand contains the mutated sequence, that is, a heteroduplex.

Step 4 *Transformation to E. coli*   The double-stranded DNA is transformed into *E. coli* cells. In *E. coli*, subsequent replication will produce some double-stranded circular DNA molecules with wild-type sequences and some molecules with mutated sequences resulting in a mixture of wild-type and mutant M13 phages.

Step 5 *Screening of phages with mutant sequence*   Bacteriophages containing either the wild-type or the mutant sequence can be distinguished from each other through hybridization screening. A radiolabelled version of the synthetic oligonucleotide used to create the mutation will bind preferentially to the mutant sequence when compared with the wild-type sequence. Therefore, bacteriophage plaques that are able to bind the oligonucleotide at high stringency should contain the mutant sequence.

The primer extension SDM procedure was widely adopted in the early 1980s. It, however, suffered from the following drawbacks.

❖ The DNA to be mutated had to be cloned into M13.

❖ The efficiency of the mutagenesis procedure was quite low. If everything went right, 50% of the clones would be mutants. But in practice, the number of mutant clones is much less than this and a large number of plaques have to be screened for getting a mutant clone.

❖ The newly synthesized DNA, as it is produced in vitro, will not be methylated, while the wild-type M13 genome already synthesized in vivo will be methylated. Since the mismatch repair systems of the *E. coli* favour the repair of non-methylated DNA, the mismatches between wild-type and mutant DNA strands are repaired in favour of a return to the wild-type sequence.

❖ The differential screening procedure to identify mutant phages is both slow and cumbersome.

Methods have been devised to increase the overall efficiency of a mutagenesis experiment by either increasing the efficiency of the mutagenesis reaction itself or by the use of bacterial strains that are less likely to degrade the newly formed mutant DNA strands. For example, defective *E. coli* cells in the *mutL*, *mutS*, and *mutH* mismatch repair systems can be used for the transformation of the hybrid DNA molecules so that the mutation cannot be repaired back to the wild-type sequence.

Oligonucleotide-directed Mutagenesis with Phosphothioate Strand Selection or Eckstein Method

An effective approach to improve the efficiency of mutagenesis could be to select the mutant DNA strand by some procedure. If the mutant strand is protected, enrichment of mutants in the pool will occur resulting in efficient mutagenesis. In Eckstein method, a nucleotide analog is allowed to be incorporated thus preventing the degradation of the newly synthesized mutant strand.

This procedure is similar to the primer extension method except that phosphosthioate nucleotides are used during the primer extension step. In phosphothioate nucleotide, the phosphate–oxygen linkage is replaced by a phosphate–sulphur linkage. If this nucleotide analog is used during the DNA synthesis in primer extension step, the DNA polymerase will take up phosphothioate into the newly synthesized strand. The incorporation of this nucleotide analog will not hamper the normal base pairing but will be resistant to cleavage by certain restriction enzymes or exonucleases. In the Eckstein method, after annealing of the mutagenesis oligonucleotide to the single-stranded M13 DNA, extension by DNA polymerase is allowed in the presence of dATP, dTTP, and dGTP, and dCTP is replaced by its phosphothioate analog. Thus the mutant DNA strand (but not the wild-type) will contain a phosphothioate for each *C* residue in the strand. The DNA duplex so formed is then cleaved with enzymes like *Pst*I, which will preferentially cleave the wild-type DNA strand as there is no phosphothioate. The nicked DNA is then removed by exonuclease, which degrades DNA from ends. The resulting DNA will, therefore, be enriched in the mutant circles, which are then transformed into *E. coli*. The mutation efficiency with this method is about 40%–60%.

Oligonucleotide-directed Mutagenesis with Dut-ung- (or Kunkel) Strand Selection

Unlike phosphothioate strand selection method, the efficiency of mutation in the Kunkel method is increased by an approach that selects against the wild-type strand. The procedure comprises the following steps.

Step 1 *DNA template generation*   The gene to be mutated is placed in an M13 phage, which is grown in a *dut-ung-E. coli* strain. The *dut* gene encodes dUTPase, the function of which is to degrade deoxyuridine triphosphate (dUTP) within the cell. The *dut* mutation results in an elevated concentration of dUTP accumulating within the cell and incorporation of uracil (U) in place of thymine (T) at some positions during DNA replication. The *ung* gene encodes uracil N-glycosylase, which normally removes uracil from DNA. Thus the M13 grown in the double mutant will have U incorporated in place of T, and this error is not removed.

Uracil residues have the same base pairing properties as T; so the incorporation of U into DNA in place of T is not mutagenic in itself. Further, uracil-containing DNA templates are not distinguishable from normal templates, and the presence of uracil in the template is not inhibitory to in vitro DNA synthesis. M13 phage DNA isolated from a *dut-ung-E. coli* strain will contain approximately 20–30 U residues per 8000 bases of its genome.

Step 2 *In vitro DNA synthesis*   The U-containing single-stranded M13 DNA is incubated with mutagenic oligonucleotide, which anneals with the target DNA. The mutagenic oligonucleotide is expected to base pair with the template except at the location of mismatch. The extension reaction is then carried out with DNA polymerase and dNTPs. The extension reaction generates a complementary DNA strand containing the desired sequence alteration, but with thymine residues in place of uracil. After the synthesis of the second DNA strand is complete, the ends are covalently joined using DNA ligase. The resulting double-stranded DNA consists of the wild-type strand that contains uracil residues and the newly synthesized strand that contains the mutant bases present in the oligonucleotide, but no uracil residues.

Step 3 *Enrichment of mutant strand*   The double-stranded DNA is then transformed into *ung*+ wild-type *E. coli* cells, where the uracil N-glycosylase recognizes the uracil residues in the DNA and excises the uracil base to leave apyrimidinic (AP) sites in the template strand. These AP sites are lethal lesions, presumably because they block DNA synthesis and are sites for the incision by AP endonucleases, which produce strand breaks. Thus the template strand is rendered biologically inactive and is degraded. Hence, when the double-stranded DNA is introduced into the *ung+ E. coli*, only the mutant strand will be replicated. Using this approach, mutation efficiencies approaching 100% can be obtained (Figure 3).

Cassette Mutagenesis

The cassette mutagenesis method can be easily applied in cases where there are two restriction enzyme recognition sites spanning the DNA fragment to be mutated. The plasmid containing the insert DNA is at first digested with the two restriction enzymes so as to cleave the desired fragment from the plasmid. It will contain two fragments—one is the large fragment representing the majority of the plasmid, while the other is a small fragment representing the DNA to be mutated. These two fragments are then separated by standard procedures of plasmid purification. A double-stranded oligonucleotide cassette is then synthesized, which contains the required mutation and the required overhanging sequences needed for the ligation to the restriction enzyme cleavage sites. During ligation, the new cassette will replace the cleaved sequences and produces mutant. This procedure is highly efficient at producing mutations, provided that the small wild-type DNA fragment can be eliminated. The drawback of the technique is that it mutates efficiently only when there are two appropriate restriction sites available spanning the region to be mutated.

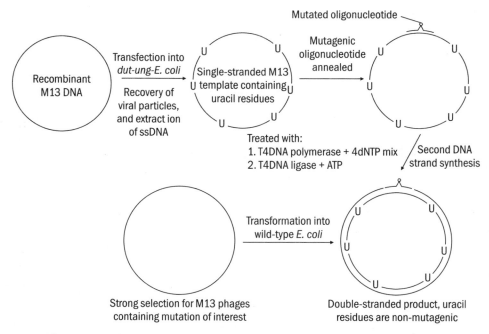

Figure 3 Oligonucleotide-directed mutagenesis with strand selection

PCR-based Mutagenesis

PCR experiments can be successful even if the oligonucleotide primers do not match exactly the target sequence. So by designing the primers appropriately, one can introduce mutations in the PCR product specifically at the ends. During a PCR reaction, the primers start the DNA replication process and are incorporated into each strand. In this way, these primer sequences are present in the final product. As a result, if any base change is inserted into the primer sequence, it will be incorporated into the amplified product. However, during PCR reaction, only the 3'-end of the PCR primer sequence is required to match the target sequence. If the 3'-end does not match, the polymerase cannot extend the primer efficiently and the reaction either fails or is inefficient. Therefore, the changes that are introduced at the 5'-end of the primer will be carried to the final product. This is utilized in inserting restriction sites into a PCR product. Hence, PCR can be a powerful tool to introduce mismatches in a linear DNA fragment. This has been utilized for creating SDM in a rapid and efficient manner.

A variety of PCR-based methods have been developed to create SDM. In the simplest case, if the site of the mutation is near the end of the gene, a primer containing mismatch can be used for the PCR reaction. However, in most cases, the desired site is not present at the end. So a different strategy has to be used, which is more complex. The following method, often referred to as the mutagenesis by overlap extension PCR, enables one to generate mutations at any site on the PCR product. This method requires four oligonucleotide primers and three separate PCR reactions, as illustrated in Plate 2.

Procedure for Mutagenesis by Overlap Extension PCR

The procedure for mutagenesis by overlap extension PCR involves the following steps.

❖ *Selection of primers* Two outermost primers (primers 1 and 4) are designed, which are complementary to the antisense strand and the sense strand of the target DNA, respectively. In addition, two internal primers—primers 2 and 3—are taken, which overlap (complementary to each other) and contain the alteration in the sequence, but in opposite strand, so that the required mutation can be introduced to each strand.

❖ *PCR reactions* Two PCR reactions are carried out in two separate tubes but with the same reaction conditions. In one reaction, the 5'-end of the gene is amplified with the primers 1 and 2. The resulting product will bear the mutation at its 3'-end. In the other PCR reaction, the 3'-end of the gene is amplified using primers 3 and 4 so that the resulting product will bear the mutation at its 5'-end. Since primers 2 and 3 overlap with each other and are complementary, the 3'-end of the former PCR product (generated with primers 1 and 2) will be identical to the 5'-end of the latter PCR product (generated with primers 3 and 4).

❖ *Denaturation and annealing of products* The products of the two reactions are then mixed, denatured, and re-annealed; some of the strands from the first PCR product will anneal with that obtained from the second reaction especially at the regions of overlap, that is, at the regions corresponding to the sequences of primers 2 and 3. Also there will be annealing of the products obtained in the first PCR as well as in the second PCR.

❖ *Overlapping extension reaction* The hybrid molecule containing a short 3′-end can act as a primer and can be extended in the next phase by DNA polymerase to form a complete double-stranded DNA molecule with mutation on both strands. The other type of hybrid molecule cannot be extended as it has overlap at the 5′-end.

❖ *Final PCR step* The subsequent and final PCR carried out with flanking primers (primers 1 and 4) will then amplify the full-length product to yield the mutated DNA.

The mutagenesis efficiency of PCR methods is very high. The amplification steps ensure that practically no wild-type DNA will be present in the final product. The early attempts to use PCR mutagenesis suffered from the drawback that some unwanted mutations were introduced due to the absence of proofreading activity in Taq DNA polymerase. The proofreading activity, which is the 3′–5′ exonuclease activity of DNA polymerase, is capable of removing any mismatched nucleotide incorporated onto the template by mistake. Since Taq DNA polymerase does not have this proofreading activity, it has a high error rate (about 1 in 9000 nucleotides). The availability of other thermostable DNA polymerases has reduced the problem. In any case, the mutated product should be checked by sequencing to ensure the specificity of mutation.

CONSIDERATIONS FOR DESIGN OF MUTAGENIC OLIGONUCLEOTIDES

Whatever be the method of mutagenesis, the success depends on the design of the oligonucleotide in all these methods since the mismatch in the target sequence is introduced through oligonucleotide primer. If these oligonucleotides are correctly designed, the mutants will be produced at ease but incorrect design may lead to low efficiency and even failure in mutagenesis. The oligonucleotide must have the following characteristics.

❖ It must be complementary to the appropriate strand of the target DNA.

❖ The length of the oligonucleotide should be appropriate for annealing. Too short sequences may not anneal, and very long sequences are difficult to handle in the reaction. Usually for a single base change, the length of the oligonucleotide is of the order of 20–25 bases. For the introduction of more complicated mutation, longer sequences are required, which may even extend up to 80 bases.

❖ The oligonucleotides must carry the mismatched bases at the centre. This will allow formation of stable hybrids with the template.

❖ The oligonucleotides must be free of palindromic, reiterated or self-complementary sequences. These sequences will form secondary structures and impair the efficiency of duplex formation and in the long run will hamper mutant formation. If such structures are inevitable, longer oligonucleotides may be constructed for the success of the method.

SUMMARY

- Site-directed mutagenesis (SDM) is a technique widely used in molecular biology for creating mutation in the DNA molecule in a defined location.

- The primer extension SDM procedure starts with a single strand of DNA template of which a short oligonucleotide containing the desired mutation is annealed to the target region. It then serves as a primer for the synthesis of the complementary DNA strand. The newly synthesized strand contains mutation. The efficiency of mutation is about 50%.

- To increase the efficiency of mutagenesis, the mutant DNA strand is selected by some procedure. Strand selection methods distinguish between the newly synthesized mutant DNA and the parental wild-type sequence.

- In the Eckstein method of strand selection, a nucleotide analog is allowed to be incorporated, which prevents the degradation of the newly synthesized mutant strand.

- The Kunkel method distinguishes the two strands on the basis of the presence of uracil. This method utilizes dut-ung-*E. coli* strains for the incorporation of uracil residues in the wild-type strand. The newly synthesized strand containing mutant bases does not contain uracil.

- Cassette mutagenesis cleaves the plasmid at a restriction site, and a double-stranded oligonucleotide cassette containing the required mutation is ligated. During ligation, the new cassette replaces the cleaved sequences and produces mutant.

- PCR can also be used for SDM by using oligonucleotide primers containing the desired mutation.

REVISION QUESTIONS

1. How can site-directed mutagenesis be achieved in a cloned cDNA sequence?
2. What type of vector is used for site-directed mutagenesis? How would you transfer the cDNA clone from its original vector to the vector used for site-directed mutagenesis?

3. The partial sequence of a protein is Met-His-Leu-Val-Pro-Gly-Val-….. Generate a DNA code for this sequence and describe in detail how you would use site-directed mutagenesis to change proline to serine.

4. Describe the cassette mutagenesis method.

5. Describe briefly the PCR method of mutagenesis and its advantages and disadvantages.

6. What are the characteristics that an oligonucleotide must possess for ensuring success of the various mutagenesis methods?

9

Restriction Fragment Length Polymorphism

OBJECTIVES

After reading this chapter, the student will be able to:
- Describe the basic RFLP procedure
- Explain the advantages and disadvantages of the RFLP technique
- Discuss applications of the RFLP technique

INTRODUCTION

The term "restriction fragment length polymorphism", or RFLP, denotes a difference in the positions of restriction sites in two or more homologous DNA molecules. The RFLP analysis was one of the first techniques to be widely used for detecting variation at the DNA sequence level. The principle behind the technology rests on the comparison of band profiles generated after restriction enzyme digestion of DNA molecules of different individuals or different strains of bacterial species and so on. The variations in DNA sequences produce restriction-digested fragments of variable lengths. These differences in fragment lengths can be seen after gel electrophoresis, hybridization, and visualization. When DNA samples from two different individuals or from two different bacterial strains are subjected to RFLP analysis, different band patterns will be obtained if there are some variations in the location of specific restriction sites (Figure 1).

There are various types of changes in a genome, which can give rise to polymorphism. These are as follows.

❖ Point mutations, which can cause a loss or gain of a specific restriction site, but this happens at a low frequency.

❖ Insertion of transposable elements or insertion sequences (or IS elements), which will make the restriction fragments longer or shorter, depending on whether the restriction site is absent or present in the inserted sequence.

❖ Deletion of small or large sequences, which might change the RFLP pattern.

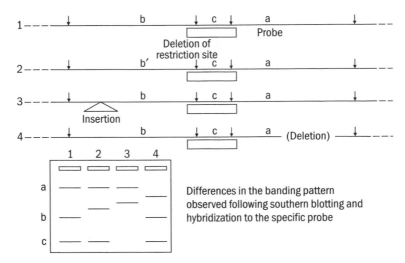

Figure 1   Restriction fragment length polymorphism

❖ Duplication or transposition of repeated sequences (short or long repeats).

❖ Genetic processes like translocations and inversions can also lead to RFLP.

BASIC PROCEDURE OF RFLP

The basic RFLP procedure involves the following steps.

Step 1 *Isolation of DNA*   Like other DNA-based technologies, isolation of DNA is the first step. To extract DNA, the cell wall and the nuclear membrane are broken by several laboratory procedures and the DNA is then appropriately separated from other cell components. While doing so, care must be taken to ensure that the process does not damage the DNA molecule and that it is recovered in the form of a long thread.

Step 2 *Digestion with restriction enzymes*   Extracted DNA is digested with specific, carefully chosen restriction enzymes. Each restriction enzyme, under appropriate conditions, will recognize and cut DNA in a predictable way, resulting in a reproducible set of DNA fragments ("restriction fragments") of different lengths. In case of genomic DNA or large DNA fragment of a genome used for restriction digestion, the number of digested fragments should be very high.

Separation of restriction fragments by electrophoresis   Millions of restriction fragments produced are commonly separated by electrophoresis on agarose gels and visualized following ethidium bromide staining. In case the number of fragments is large, these would be seen as a continuous "smear". So mere staining of gels cannot detect the polymorphism. Hybridization must, therefore, be used to detect specific fragments.

Step 3 *Hybridization*   Southern hybridization is used. The gel is first denatured in an alkaline solution and placed in a tray. A porous nylon or nitrocellulose membrane is laid over the gel and the whole weighted down. All DNA restriction fragments in the gel transfer as single strands by capillary action to the membrane. All fragments retain the same pattern on the membrane as on the gel.

The membrane with the target DNA is then incubated with the desired DNA probe, which is converted into a single-stranded molecule and labelled by either a radioisotope or a chemical compound such as digoxygenin. Incubation conditions are such that if strands on the membrane are complementary to those of the probe, hybridization will occur and labelled duplexes will be formed. Usually, conditions are chosen in such a way that non-specific hybridization is minimized (stringent conditions). Thus the DNA probe picks up homologous sequences among thousands or millions of undetected fragments that migrate through the gel. Visualization of fragments is done after the exposure of the hybridized membrane and an X-ray or photographic film.

Step 4 *DNA probe*   A subset of all restriction fragments generated is detected by the use of DNA probe. The probe may be chosen from a genomic DNA library, from cDNA library or a previously known gene, which may be generated by polymerase chain reaction (PCR). To discriminate between different individuals, repetitive sequences are widely used as probes of RFLP analysis. They may be minisatellites containing repeated sequences in tandem (that is, head to tail) and occur at many loci on the genome. This type of RFLP with repeated sequences as probes has wide application in DNA fingerprinting in forensics.

ADVANTAGES AND DISADVANTAGES

The advantages of the RFLP technique are as follows.

❖ Simple and robust.

❖ Sequence information of DNA is not required.

❖ Can be applied at the species level or individual level.
   The disadvantages are as follows.

❖ Large amount of DNA is required.

❖ The assay is time consuming, and only a few loci are detected per assay.

❖ Low level of polymorphisms is found in some species; so this technique will not be suitable for those.

❖ Choice of probe is a critical parameter.

APPLICATIONS

The RFLP technique has widespread applications, some of which are discussed as follows.

❖ *Physical mapping of genomes*   The RFLP technique has also been used for physical mapping of genomes, wherein the fragments are separated by restriction enzymes like *Not*I, I-*Ceu*I, *Sfi*I, which cut the genomes less frequently and are then separated by pulsed-field gel electrophoresis and probed.

❖ *Strain identification and in epidemiology of bacterial diseases*   Each bacterial species can be identified by its distinct RFLP profile using a suitable probe. Here usually some conserved genes like those coding for ribosomal proteins or often rRNA genes are used for profiling.

❖ *Diagnosis of genetic diseases*   These include cystic fibrosis, Huntington's chorea, and sickle-cell anaemia. In particular, sickle-cell anaemia is caused by a single mutation of a single nucleotide; thymine is replaced by adenine. This mutation occurs at a point in the DNA sequence recognized by the restriction enzyme *Mst*II in a person without the disease. The RFLP from a person suffering from sickle-cell anaemia will have a long band instead of two shorter ones because the cleavage by *Mst*II will not occur (Figure 2).

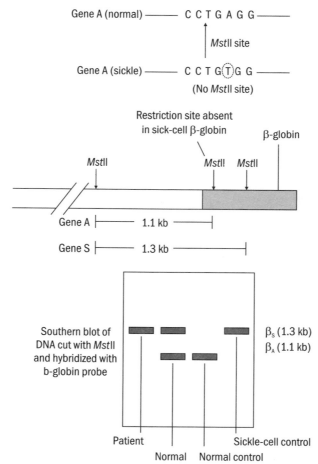

Figure 2   Diagnosis of sickle-cell anaemia by RFLP

Besides, RFLP is also used in the analysis of genetic diversity, genotyping, forensics, paternity tests, origin and evolution of species, localization of genes for genetic disorders, and determination of risk for disease.

SUMMARY

- The restriction fragment length polymorphism (RFLP) technology detects length changes in target DNA molecules following restriction enzyme digestion.
- The target DNA is hybridized with an appropriate DNA probe to generate banding pattern.
- RFLP banding patterns reflect variations in nucleotide sequences at the hybridization site of the probe or its neighbouring region.
- RFLP is a highly robust technology and has applications in genetic diagnosis and various other fields.

REVISION QUESTIONS

1. Define RFLP and describe the procedure used to detect RFLP.
2. Why is nucleic acid hybridization required for the detection of RFLP?
3. How can RFLP markers serve as genetic markers?
4. Suppose one is detecting RFLP patterns of four different bacterial genomes. They have all been digested by *Bam*HI, hybridized a probe A, and a banding pattern is obtained. What will happen if all of these DNAs are digested with another restriction enzyme and hybridized with the same probe?
5. Describe the advantages and disadvantages of the RFLP technique?

10

Polymerase Chain Reaction

OBJECTIVES

After reading this chapter, the student will be able to:
- Describe the basic reaction method
- Explain the essential components of PCR
- Understand hot start protocol
- Discuss nested, inverse, and anchored PCR
- Explain reverse transcription polymerase chain reaction
- Understand real-time PCR
- List the applications of PCR

INTRODUCTION

The polymerase chain reaction (PCR) is the most powerful technique that has been developed in recent times and has revolutionized the area of recombinant DNA research. In this technique, a specific DNA fragment can be amplified a billion fold in a few hours' time. The reaction is simple and can be easily performed. Unlike the amplification of DNA using cloning and propagation into a host cell, PCR amplification is carried out entirely in vitro.

THE BASIC REACTION METHOD

PCR is essentially a DNA polymerase reaction. Like any other DNA polymerase reaction, it requires a DNA template and a free 3'-OH group to get the reaction started. The template is the DNA molecule itself, which is present in small or sometimes minute quantity. The 3'-OH groups at both ends of the target to be amplified are provided by two oligonucleotide primers, which are complementary to the two ends of the target. In vivo DNA polymerase requires an RNA primer to start the replication reaction; in vitro, a more easily synthesized DNA oligonucleotide serves the purpose. The PCR amplifies the DNA using repeated cycles, each consisting of essentially three steps, which are denaturation, annealing, and extension.

The reaction mixture contains the DNA molecule to be amplified, thermostable polymerase to carry out DNA synthesis, nucleotide triphosphates (dNTPs) as the building blocks, and two primers for starting DNA synthesis, suitable reaction buffer, Mg^{2+}, and so on. The basic steps of the PCR reaction are as follows.

Step 1 *Denaturation* The reaction mixture is heated to 95°C when hydrogen bonds break and the two strands of DNA separate, a process normally called denaturation or the melting of DNA. We know that DNA can be reversibly denatured by a cycle of heating and cooling.

Step 2 *Annealing* The temperature is lowered to about 55°C (depending on the primer used). In this step, the primer binds to the DNA, a process known as annealing or hybridization. One of the primers recognizes and binds to one strand, and the other primer recognizes the other strand. The two primers, termed forward and reverse primers, are designed in such a way that the free 3'-end of each primer faces the other one so that the DNA synthesis proceeds on both the strands.

Step 3 *Extension* The temperature is increased to 72°C, which is the ideal working temperature for the thermostable polymerase. The polymerase at this step adds nucleotides to the growing DNA strand in a 5' to 3' direction.

Each time the above steps are repeated, the number of copies of the target molecule (DNA sequence between two primers at the starting DNA molecule) becomes double. After 20–25 cycles, a million of DNA molecules are formed from a single segment of double-stranded DNA.

A typical PCR product, when visualized on an agarose gel, yields a clear single discrete band. This suggests that the DNA fragments produced are homogenous and that they begin and end at the same point. The reaction is clearly illustrated in Plate 1 where the products formed during the first three cycles are shown.

In cycle 1, the two target DNA strands are separated, and DNA replication initiates at the point of primer binding. At the end of cycle 1, the two newly synthesized DNA strands each will have defined 5'-ends, but 3'-ends are ill-defined. DNA synthesis will not terminate at a specific point, but will only stop during the heat denaturation step of cycle 2. Each of the DNA strands present at the end of cycle 1 will proceed to the next cycle of PCR and each will act as a template for the binding of new primers. In cycle 2, primer binding to both the original template strands and the strands synthesized during cycle 1 will occur. Primer binding to the original template strands will result in the formation of the same products that were made during cycle 1. However, primer binding to the DNA strands produced in cycle 1, followed by replication, will result in the formation of a DNA strand with both defined 5'-end and defined 3'-end. This occurs because DNA replication will terminate when there is no more DNA sequence to copy. Thus at the end of cycle 2, two DNA strands are formed that have a defined 5′-end and a defined 3′-end. These are, however, base paired to DNA fragments that have ill-defined 3′-ends. Again the products from cycle 2 of the PCR process will go forward into cycle 3, and each DNA strand again can be used as a template for primer binding. At the end of cycle 3, two double-stranded DNA molecules are formed that have 5′- and 3′-ends beginning and ending at the positions of primer binding within the original target DNA sequence.

Beyond cycle 3 of the PCR, the repeating cycles of heat denaturation, primer annealing, and extension will result in the exponential accumulation of specific target fragments of DNA.

In the original experiment of PCR reaction conducted by Kary Mullis in 1985, *Escherichia coli* DNA polymerase was used. Since *E. coli* DNA polymerase is sensitive to heat, its activity was drastically reduced following heat denaturation at 95°C and fresh enzyme had to be added during each cycle. The discovery of Taq DNA polymerase from the thermophilic bacterium *Thermus aquaticus* made a breakthrough. The purification of Taq DNA polymerase (and later cloning and expression in *E. coli*) revealed that this polymerase is stable at high temperature, and the PCR reaction became much simpler. Manual addition of enzyme at each cycle was not required. At the same time, the extension reaction could be done at a higher temperature compared to the annealing reaction, which means that specificity of annealing is not compromised. The PCR procedure was automated following the development of thermal cycling machine or thermal cyclers. These instruments are capable of rapidly switching between the different temperatures that are required for the PCR reaction. Thus the assembled reactions are placed in a heating block with suitable thermal cycling programme.

ESSENTIAL COMPONENTS OF PCR

PCR relies on some key chemical components such as thermostable DNA polymerase, small amount of DNA that serve as the initial template or target sequence, a pair of primers that bind to the target sequence, four deoxyribonucleotides (dNTPs), and a few ions and salts. The PCR reaction uses these ingredients to mimic the natural DNA replication process. A machine called thermocycler is used to automate the PCR process. The thermocycler has the capability to "jump-start" each stage of the reaction by raising or lowering the temperature of the PCR reaction mixture at preset conditions. The components required for PCR reaction are described below.

Thermostable DNA Polymerase

For routine PCR, Taq DNA polymerase is used. Taq DNA polymerase is a monomeric enzyme isolated from the bacterium *T. aquaticus*, which is a hyper thermophilic organism. It is a monomeric enzyme of 94 kDa. The enzyme is thermostable. It replicates

DNA at 74°C and remains functional even after incubation at the denaturing temperature of 95°C.

The specific activity of the most commercial preparations is about 80 000 units per milligram of protein. For a normal PCR reaction, 0.2–0.5 units for 25–50 µl reaction are sufficient.

The Taq DNA polymerase possesses a 5′–3′ polymerase activity and a 5′– 3′ exonuclease activity, but it lacks 3′–5′ proofreading activity. Thus if an incorrect base is incorporated into the growing polynucleotide chain, it cannot be removed, and as a result, some errors crop up in the amplification process due to mis-incorporation of nucleotides. In vitro assays suggest that Taq mis-incorporates one base in every 10^4 to 10^5 bases. In certain applications, especially when the PCR product is cloned, it is desirable to check the clone by sequencing for any changes in the DNA sequence that may have occurred during amplification. The fidelity of the amplification reaction can be assessed by cloning, sequencing, and comparing several independently amplified molecules.

Another characteristic of Taq DNA polymerase is that it has a tendency to incorporate a deoxynucleotide (often an adenosine) in a template-independent manner on the 3′-end of the newly synthesized DNA strand. The PCR products produced by Taq do not have blunt ends but have a single 3′-A residue overhang. This property has been exploited to aid the cloning of PCR products.

Since the discovery of Taq DNA polymerase, a number of other thermostable DNA polymerases have been discovered and used in PCR experiments. Although Taq remains the most widely used enzyme for PCR, other enzymes are suitable for some specific applications. These thermostable enzymes differ in fidelity, efficiency, and ability to synthesize long fragments. They are now commercially available for various applications.

Primers

The efficiency and specificity of the amplification reaction are crucially dependent on primers. The primers need to be designed carefully to obtain products with high yield (efficiency), to minimize amplification of unwanted sequence (specificity), and also for subsequent studies. The design of primers has remained qualitative and empirical. The following factors are important for choosing effective primers.

❖ The optimum length of primers should be between 17 and 30 nucleotides, which is sufficient to allow annealing to a single sequence in a genome.

❖ A GC content of about 50% is ideal as the choice of annealing temperature depends on the melting temperature (T_m) of the primer, which in turn is dependent on the GC content (Box 1). For primers with a low GC content, a long primer is chosen to avoid a low melting temperature.

Box 1 Calculation of melting temperature of hybrids between oligonucleotide primers and their target sequences

Several equations are available based on the theory of hybridization. None of these could be 100% accurate; so the choice is entirely on the person performing experiment. However, the same equation should be used for each primer of a pair. The melting temperature or T_m may be calculated from the following equation.

$$T_m = 81.5 + 16.6 \log [K^+] + 0.41 \ (GC\%) - 675/n$$

where n is the number of bases in the oligonucleotide and K^+ is the cationic concentration in the solution. This equation is derived originally by Bolton and McCarthy (1962) and later modified by Baldino *et al.* (1989).

Another empirical equation is commonly used in laboratories and is known as "the Wallace rule" and stands as follows.

$$T_m = 2(A + T) + 4(G + C)$$

where $A + T$ is the sum of A and T residues and $G + C$ is the sum of G and C residues in the oligonucleotide. This relation is most effective for oligonucleotides 15–20 mer in length and at cation concentration equivalent to 1 M NaCl.

These equations do not consider the effect of neighbouring bases and are independent of the sequence per se. However, algorithms have been developed incorporating nearest neighbour thermodynamic parameters into the equation and appear to predict the melting temperature more accurately. For most purposes, the Wallace rule and the Baldino equation are adequate and are widely used as these are simple to apply.

❖ Sequences with single nucleotide tracts should be avoided to prevent binding of the primer to repetitive sequences in the genomic DNA.

❖ Primers should not have considerable secondary structures; for example, the presence of palindromic sequence within individual primers will form hairpin structures.

❖ Complementarity between the two primers should be avoided.

Primers designed considering the above guiding principles may be assumed to work, which is based on common sense and experience. A typical example is shown in Figure 1.

In some applications, it becomes desirable to amplify several segments of the template DNA by different sets of primers in the same reaction tube. These reactions are termed "multiplex PCR", and the choice of primer concentration is crucial to avoid non-specific reactions.

Oligonucleotide primers are synthesized in automatic DNA synthesizer. Such primers can be directly used for simple PCR reactions, but purification of the oligonucleotides by chromatography on commercially available resins or by denaturing polyacrylamide gel electrophoresis increases the efficiency of PCR reactions.

GTGAACACGGGCAGGGGAAGAAGGGCCAGTGATTGTCTCACCCCAAACTCCTCCCACCCA

CCCGTACCAAGGTTCCTCTTCTCTGACTCTGCAAACCTCTGGGAAGCTCTCAGGCCTCAC

AGAGAAGACTGGAGGGCCGGATCTGGTTCACTCTTTCCCTAGGGTCTGAAGGATCTGCGC

CCTGTGTTCAAGGACGACGCTGAAACATTCTTTTCTGCGTGCCTGATCTGAGGCCTCACA

ACTCAGGTGAGCAAGACTGTGTCCCTTTCACATGGAGAAGCCGAGTTCCACGGAGGTTGA

GTGAGGAGCCGGCATCACACAAGGACTCGGAGGGGGCTCCTCGAGACTCGCGGGACAGAGA

CGGCAGCCCCAGGTGGGAAAGGCCCAGGCCAGGGACACGCTGGGCCTGGGGTGGGGGAAG

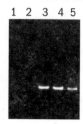

Figure 1 Primer design and PCR product

Computer-assisted Design of Primers

Many computer programs are available for design of PCR primers. Such programs often generate a number of primers and rank the potential primers according to their efficiency. T_m is calculated by using the nearest-neighbour method in which the thermodynamic stability of the primer–template duplex is derived from the sum of interactions of neighbouring bases.

Deoxynucleotide Triphosphates

A standard PCR reaction contains equimolar amounts of each dATP, dCTP, dGTP, and dTTP. dNTPs specifically made for PCR are free of pyrophosphates, which may inhibit PCR reaction. Normally, dNTPs are stored in aliquots at –20°C.

Divalent Cations

All thermostable DNA polymerases require divalent cations, especially Mg^{2+} for their activity. In a PCR reaction, the concentration of Mg is very important as at low concentrations of Mg, the reaction fails and at very high concentration, the reaction loses its specificity. So the optimum concentration has to be worked out for each reaction. Generally, 1–5 mM concentration is employed.

Mg^{2+} binds to the oligonucleotide primers as well as dNTPs. The molar concentration of the cation must exceed the molar concentration of phosphates contributed by dNTPs and primers. The optimum concentration will depend on the primer and dNTP concentrations. Further, being a cofactor in the reaction, increased Mg^{2+} concentrations are often effective in reducing non-specific PCR products.

Buffers

PCR buffer usually contains 10 mM Tris-HCl adjusted to pH 8.3–8.8 at room temperature. Standard PCR buffer also contains monovalent cations to improve the yield of the product.

Template DNA

The template DNA concentration should be low for increased efficiency of a PCR reaction. Typical amounts used for yeast, bacterial, and plasmid DNAs in a PCR reaction are of the order of 10 ng, 1 ng, and 1 pg, respectively. Compared to linear templates, covalently closed circular DNAs are amplified with slightly less efficiency. The length of the target sequence is not important; however, in case of large DNA template (>10 kb), digestion with restriction endonuclease, which does not have recognition site at the target region to be amplified, is often recommended.

Purity of template is important as any contaminating template might be amplified. Also the presence of minute amount of proteinase *K*, detergent, phenol, and ethylenediaminetetraacetic acid (EDTA) might inhibit the reaction. The DNA template used can be cleaned up by dialysis, ethanol precipitation, extraction with chloroform or chromatography through a suitable resin.

Cycling Temperature and Duration

A typical cycling condition is like this: 95°C, 30 s denaturation; 60°C, 30 s annealing; and 72°C, 1 min extension. The denaturing and annealing steps are relatively short, but are sufficient to break and reform the hydrogen bonds between DNA strands. Longer denaturation step may introduce nicks in the template DNA. Depending on the length of the extended product, the time of extension is selected. The annealing temperature depends on the melting temperature, or T_m, of the primers used. If T_m is lower, the annealing temperature has to be lowered, but lowering of temperature reduces the probability of specificity.

The number of PCR cycles performed depends on the amount of the initial DNA template in the reaction and on the amount of DNA required after the amplification process. To avoid replication errors, the number of cycles should be as few as possible; normally 20–25 cycles are carried out. After the completion of cycles, a final extension step is included to ensure that the entire DNA in the reaction has been replicated into a double-stranded product.

Hot Start Protocol

Taq DNA polymerase has a 5' to 3' exonuclease activity, which means that the enzyme is able to degrade the oligonucleotide primers within the PCR reaction. This is particularly relevant during the first denaturing step of cycle 1, when the oligonucleotides are not bound to the DNA template and the polymerase is free in solution. During the first heating cycle, the temperature of the PCR mix rises from room temperature to 95°C. Under this condition, the temperature within the tube would be 72°C, which would be optimum for the polymerase. Since none of the oligonucleotides are bound to the template DNA, the enzyme will be unable to replicate the DNA and will tend to result in primer degradation and subsequent inefficient PCR. Also there could be a second possibility. At temperatures below the desired hybridization temperature for the primer (typically in the region 45–60°C), mismatched primers may form and may be extended somewhat by the polymerase. Once extended, the mismatched primer is stabilized at the unintended position. Having been incorporated into the extended DNA during the first cycle, the primer will hybridize efficiently in subsequent cycles and, hence, may cause the amplification of a spurious product. To overcome these problems and to prevent non-specific PCR products from being synthesized prior to cycling, Taq DNA polymerase can be added to the reaction mix already at 95°C, that is, after the heat denaturation step of the first cycle. Such a procedure is called "hot start", which increases both the yield and specificity of the PCR reaction.

Alternatives to Hot Start

Taq DNA polymerase can be mixed with a specific antibody that binds to the enzyme and inhibits its activity at low temperature but is inactivated at the denaturation step.

❖ AmpliTaq GoldTM, a modified Taq polymerase that is inactive until heated to 95°C.

❖ SELEX (systematic evolution of ligands by exponential enrichment) method in which the polymerase is reversibly inactivated by the binding of nanomolar amounts of a 70 mer, which is itself a poor polymerase substrate and should not interfere with the amplification primers.

Nested PCR

Nested PCR is a modification of the conventional PCR to minimize the formation of spurious products. In normal PCR, primers are designed complementary to the target DNA. A common problem is that the primer may bind to the incorrect region of the DNA and amplify to give spurious products. The nested PCR strategy employs two pairs of primers to amplify a specific target. The first pair of primers is complementary to the two ends of the target, and the second pair is intended to amplify an internal fragment of the target sequence. In the first run, the target DNA is amplified, which then undergoes a second run with the internal primers. It is unlikely that a spurious product would have matching internal sites as well; so the product from the second run is expected to be free of contamination due to primer dimer, hairpins, and alternative primer sequences.

Inverse PCR

Inverse PCR is a variation of PCR to amplify DNA of unknown sequence if a small region of known sequence is present nearby. The strategy is illustrated in Figure 2. A target sequence is identified with a region of known sequence. The target DNA is cut into smaller fragments with restriction endonuclease to generate a few kilobases of fragments. Self-ligation is induced at low concentrations so that circular products are obtained. A known restriction endonuclease is then employed, which cleaves the internal known sequence. This generates a linear product with known terminal sequences. PCR primers can now be designed complementary to the two ends of the linear product, which will result in the amplification of the unknown sequence as well.

Inverse PCR is applied to amplify and clone sequences flanking a known sequence (Figure 2).

Anchored PCR

Anchored PCR is a strategy employed when the sequence of only one end of a target is known, or in other words, only one primer can be designed. In this technique,

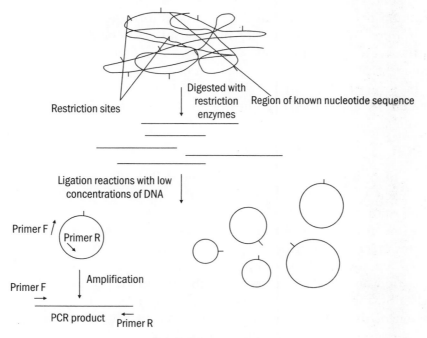

Figure 2 Inverse PCR

one primer is used to synthesize only one strand at first, after which a poly(G) tail is attached at the end of the newly synthesized strand. The newly synthesized strand with poly(G) at 3'-end will become a template for the daughter strand synthesis using an anchored primer with which a poly(C) sequence is linked. Next cycle onwards, both the anchored primer and the original primer will be used for amplification in a conventional manner.

Reverse Transcription–Polymerase Chain Reaction

Reverse transcription–PCR (RT–PCR) serves as a method of RNA amplification and quantitation after its conversion to DNA. RT–PCR can be used for complementary DNA (cDNA) cloning, cDNA library construction, and probe synthesis. In RT–PCR, the RNA strand is first reverse transcribed into the cDNA using the enzyme reverse transcriptase and is then subsequently amplified by PCR, as illustrated in Plate 2.

For cDNA synthesis, the RT reaction uses an RNA template (total RNA or mRNA), a primer (random or oligo dT primers), dNTPs, buffer, and a reverse transcriptase enzyme. This reaction generates a single-stranded DNA molecule complementary to the RNA (cDNA). With cDNA as a template, the PCR reaction is carried out. During the first cycle of PCR, the single DNA strand is made double stranded through the binding of another complementary primer, and the action of Taq DNA polymerase. RT–PCR provides the information on gene expression. Like standard PCR reaction, the products are visualized following gel electrophoresis and staining by ethidium bromide. Since PCR amplification is exponential, large differences in target concentration (100-fold or more) may produce the same intensity of band after 25 or 30 PCR cycles. Therefore, RT–PCR requires careful optimization when used for quantitative mRNA analysis and are often called semi-quantitative. Usual practice is to amplify a housekeeping gene along with the target in the same PCR tube, so that the relative amount of the target can be estimated by densitometric scanning (measuring the relative intensity of the two bands in the gel).

Real-time PCR

Real-time PCR, also called quantitative real-time PCR (Q-PCR, or qPCR), is a technique based on PCR used to amplify and simultaneously quantify a target sequence. Real-time PCR combined with reverse transcription may be used to quantify mRNA in cells and tissues. Real-time reverse transcription PCR is denoted by qRT–PCR. The acronym RT–PCR denotes reverse transcription PCR and not real-time PCR. Quantitative real-time RT–PCR, based on reverse transcription and fluorophores, permits measurement of DNA amplification during PCR in real time; the amplified product can be monitored at each cycle. Usually, the data generated is fed to a computer, which calculates the relative gene expression or mRNA copy number. Real-time PCR can also be applied to DNA samples to determine the presence and abundance of a particular DNA sample.

Real-time PCR systems use a fluorescent reporter, the signal of which is directly proportional to the amount of PCR product in a reaction. In the simplest form, a double-stranded DNA-specific dye SYBR Green is used. It binds double-stranded DNA, probably in the minor groove and fluoresces upon excitation. When the dye is included in a PCR reaction as a PCR product accumulates, the fluorescence increases, which is monitored in the screen and quantitated by computer software. Advantages of SYBR Green are that it is inexpensive, easy to use, and sensitive. The disadvantage is that SYBR Green will bind to any double-stranded DNA in the reaction, including primer dimers and other non-specific reaction products, which can result in an overestimation of the target concentration. For general purpose, SYBR Green works fine as spurious products are often generated during late cycles.

The alternative method for quantifying PCR products is using TaqMan technology, which relies on fluorescence resonance energy transfer (FRET) of hybridization probes for quantitation (Plate 3). Fluorescent reporter probes detect only the DNA containing the probe sequence and, therefore, raise the specificity and enable quantification even in the presence of spurious products. TaqMan probes are DNA sequences with a reporter dye at one end (5') and a quenching dye at the opposite end (3'). The probe is designed to hybridize to an internal region of a PCR product. The close proximity of the reporter to the quencher prevents fluorescence. Breakdown of the probe by the 5' to 3' exonuclease activity of the Taq DNA polymerase breaks the reporter quencher proximity and thus allows unquenched emission of fluorescence that can be detected after excitation with a laser. Fluorescence increases in each PCR cycle, proportional to the rate of probe cleavage, and is measured in a modified thermocycler.

Real-time PCR is a powerful quantitative tool as it is accurate, precise, with high throughput, and relatively easy to perform. However, the cost of reagents and equipment is much higher than that of standard PCR experiments.

APPLICATIONS OF PCR

PCR has revolutionized molecular biology by allowing the amplification of minute amounts of nucleic acids. It has found applications in various fields of biology and medicine; the applications are too many to enumerate. For example, it is used in medicine, where it is applied in the identification of mutations within small amounts of human DNA; in forensic science for DNA fingerprinting and DNA typing from very low amount of DNA obtained from the few cells attained by touching a surface; by pathologists who routinely need to detect and characterize small amounts of infectious microorganisms; as well as in molecular biology laboratory and other fields. A few of the major applications of PCR are listed as follows.

Molecular Cloning

PCR has several advantages over the traditional methods of gene cloning. These are better efficiency, low amount of starting material, minimal technical skills, less time consumption, and so on.

DNA Sequencing

It is now routinely used in automated DNA sequencing and has been described in Chapter 6 (Sequencing by Sanger's Method).

Genome Analysis

Gene library construction can be done by PCR, and PCR has tremendous applications in genome sequencing projects.

Archaeology

PCR can be applied on archival DNA samples from bone and other dried biological materials and, therefore, finds application in archaeology.

Forensics

PCR finds use in forensic science as it is possible to amplify a single molecule of DNA available from any source; for example, blood stains, hair, semen, and so on. Even technology today can type DNA obtained from a touched material.

Clinical Diagnosis

Clinical diagnosis includes diagnosis of viral infection, whether it is HIV or influenza virus H1N1; diagnosis of bacterial infections, tuberculosis, enteric infections, and so on; parasitic infection like leishmaniasis; prenatal diagnosis of inherited diseases like sickle-cell anaemia, beta-thalassemia, and so on; diagnosis of some types of cancer, for example, cervical cancer caused by human papilloma virus characterizing unknown mutations in humans; assessment of infertility; sex determination of embryos; and diagnosis of some neurological disorders.

SUMMARY

- Polymerase chain reaction (PCR) amplifies a region of DNA between two predetermined sites.
- PCR requires two primers—one that is complementary to each strand of DNA and the other a DNA polymerase.
- Each cycle of PCR doubles the number of copies of the amplified DNA, and in this way, large quantities of replicated DNA are obtained.

- Inverse PCR is a variation of PCR to amplify DNA of unknown sequence if a small region of known sequence is present nearby.
- Reverse transcriptase (RT)–PCR serves as a method of RNA amplification and quantitation after its conversion to DNA.
- Real-time PCR, also called quantitative real-time PCR (Q-PCR, or qPCR), is a technique based on PCR used to amplify and simultaneously quantify a target sequence.

REVISION QUESTIONS

1. Outline the PCR method for amplifying a given stretch of DNA. Describe the components used in a PCR reaction.

2. Suppose your laboratory stock of Taq polymerase is exhausted. However, you have DNA polymerase in your stock. Can you manage to carry out a PCR reaction? If so, how would you design your experiment?

3. Explain the use of primers in PCR, clarifying why two different primers must be used. What relationship must exist between the two primers used in PCR?

4. A successful PCR experiment often depends on designing the correct primers. One of the requirements states that the T_m for each primer should be approximately the same. What is the basis of this requirement? What will happen if the T_m of the primers chosen is very low?

5. Why was the discovery of a thermostable DNA polymerase (for example, Taq polymerase) so important for the development of PCR? Where is Taq polymerase found naturally?

6. One of the ways of determining whether a sequence has been successfully amplified by PCR is to do electrophoresis and look for a band of DNA of sharply defined size. Explain how such a band arises during PCR?

7. PCR is typically used for amplification of a DNA fragment that lies between two known sequences. If you want to explore DNA on both sides of a single known sequence, what type of variation in the PCR protocol is needed to achieve this?

8. What is the difference between reverse transcriptase PCR (RT–PCR) and standard PCR? For what purpose would you use RT-PCR?

11

DNA Fingerprinting

OBJECTIVES

After reading this chapter, the student will be able to:
- Understand the technique of DNA fingerprinting
- Describe a comprehensive example of DNA fingerprinting
- Explain RFLP analysis
- Describe PCR method
- Discuss amplified fragment length polymorphism
- Explain STR analysis
- Understand the family relationship test based on DNA

INTRODUCTION

DNA fingerprinting is a technique employed in forensic science to assist in the identification of individuals on the basis of their respective DNA band pattern generated following hybridization to a probe. About 99.9% of human DNA sequences are the same in every person, but the remaining is enough to distinguish one person from another. DNA fingerprinting uses tandem-repetitive regions of DNA (called minisatellites), which are scattered throughout the human genome (Box 1). These sequences do not contribute to the function of a gene and are polymorphic or highly variable and are further named as variable number of tandem repeats (VNTRs). The term "polymorphism" describes the existence of different forms within a population; here it represents the difference in the number of repeat units. As there are many different VNTRs on different chromosomes, it is extremely unlikely that any two persons will have exactly same VNTRs. These variable regions are used to generate a DNA profile of an individual using samples from blood, bone, hair, and other body tissues and products. The profile depends on the fact that no two people, except identical twins, have exactly the same DNA sequence and that although only limited segments of a person's DNA are scrutinized in the procedure, these segments will be statistically unique.

Box 1   Repetitive DNA

DNA consisting of short base sequences repeated several times within a genome of an organism is known as repetitive DNA. It is common in all eukaryotes and can be divided into various types: satellites, minisatellites, and microsatellites. They consist of very short sequences repeated several times and occur in large clusters. A significant proportion of these repeats is of uncertain function and may be "junk", or selfish DNA. In some cases, the region containing repeated sequences has a base composition distinctly different from the base composition of the genome average. In buoyancy density gradient, it forms a fraction separate from the genomic DNA, and hence it was first called satellite DNA. Repeats of short DNA sequences, typically less than 10 base pairs, flank the centromeres of each chromosome, stretching for hundreds of kilobases along either arm of the chromosome, forming centromeric heterochromatin. Tandemly repeated, short sequences also occur at each chromosome tip (telomeric DNA). Both types are important for maintaining chromosome structure.

Satellites

Most satellites are located in the centromere region. The size of a satellite DNA ranges from 100 kb to over 1 Mb. In humans, the alphoid DNA located at the centromere of all chromosomes is an example of satellite DNA.

Minisatellites

Minisatellites are a family of genetic loci consisting of short (15–100 base pairs) sequences of DNA repeated in tandem arrays. The alleles for any particular locus have the same sequence but differ as to how many times the sequence is repeated. The size of a minisatellite ranges from 1 kb to 20 kb. One type of minisatellite is called variable number of tandem repeats (VNTRs). They are located in non-coding regions. Their repeat unit ranges from 9 base pairs to 80 base pairs. The number of repeats for a given minisatellite may differ between individuals. This feature is the basis of DNA fingerprinting. Another type of minisatellite is the telomere. In a human germ cell, the size of a telomere is about 15 kb. In an aging somatic cell, the telomere is shorter. The telomere contains tandemly repeated sequence GGGTTA.

Microsatellites

Microsatellites are DNAs consisting of repetitions of extremely short units (<10 bp). These are also known as short tandem repeats (STR); the whole repetitive region spans less than 150 base pairs. Similar to minisatellites, the number of repeats for a given microsatellite may differ between individuals. Therefore, microsatellites can also be used for DNA fingerprinting. In addition, both microsatellite and minisatellite patterns can provide information about paternity.

Tandemly repeated sequences are liable to undergo misalignments during chromosome pairing. As a result, the sizes are polymorphic and wide variations are noted among individuals. Microsatellites undergo intra-strand mispairing, when slippage due to replication leads to the expansion of the repeat.

The DNA profiling technique was first reported in 1984 by Sir Alec Jeffreys at the University of Leicester in England. Jeffreys showed that each individual has a unique pattern of these repeats; the only exception being multiple individuals from a single zygote (for example, identical twins). Since then there was tremendous advancement in the area of forensic science. The group of researchers at the University of Leicester, England, cloned human minisatellite DNA fragment containing tandem repeats of closely related variants of a short consensus sequence. They took human DNA samples digested with restriction enzymes, and the minisatellite clone (radioactively labelled) was hybridized to the digested DNA. Following autoradiography, they discovered that the DNA banding patterns were completely specific to each person. They first used a 33 base-pair repeat sequence and repeated four times within the human myoglobin gene. The pattern of unrelated persons showing the same DNA profile was very rare. This technology was made commercial in 1987.

ILLUSTRATIVE EXAMPLE OF DNA FINGERPRINTING

The typical banding pattern for DNA fingerprinting is illustrated in Figure 1. The VNTRs in the chromosome of five persons A–E are presented in 1a; the corresponding patterns obtained after gel electrophoresis (if these VNTRs are amplified by polymerase chain reaction [PCR]) or southern hybridization (if analysed by restriction fragment length polymorphism [RFLP]) are shown in Figure 1b. The difference is clear: the DNA fragments with higher number of repeats migrate slowly and, therefore, can be distinctly separated from those with lower number of repeats.

The following properties of minisatellite probes make them useful for DNA fingerprinting. These are as follows.

❖ *Constancy and reproducibility*   There is no difference in patterns generated among fresh, frozen or dried DNA samples. Even DNA from any source, be it blood, hair, or bone, yields the same fingerprint. Constant patterns are also observed when DNA is analysed after extraction from the same subject at different times.

❖ *Inheritance of patterns*   Experiments have shown that bands are inherited in a simple Mendelian fashion and are co-dominant, meaning that both the paternally and

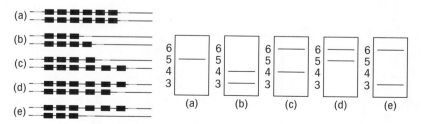

Figure 1   An example of DNA fingerprinting

maternally derived bands are distinguishable; the presence of one does not mask the presence of the other.

❖ *Individual specificity* In case of VNTRs, the mean probability of chance of match between two individuals is very little. The one exception is monozygotic twins. They have identical genotypes and thus identical DNA fingerprints.

Since the discovery of DNA fingerprinting in 1985, tremendous progress has been made in the methodology of comparing DNA samples. The main types of DNA fingerprinting methods used in forensic science are RFLP, PCR, amplified fragment length polymorphism (AFLP), and short tandem repeat (STR).

THE PROCEDURE

The first step is to isolate DNA from the available sample. Since the sample available in forensic applications is much less, some sophisticated techniques have been developed to isolate small amount of DNAs from these specimens. For creating DNA profile or fingerprinting, one of the above four methods is used.

RFLP Analysis

RFLP procedure has been described in Chapter 9. This method was used during the discovery of DNA fingerprinting. In this method, genomic DNA is digested with restriction enzymes and then electrophoresed on an agarose gel. The fragments produced are transferred to a nylon membrane and probed with a radiolabelled sequence of DNA that matches with the VNTR sequence. The VNTR fragments migrate on the gel, according to their size, and generate a pattern. The radiolabelled probe produces dark bands on the X-ray film when exposed in a time- and dose-dependent manner (Figure 2).

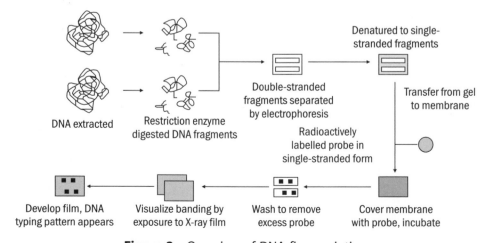

Figure 2 Overview of DNA fingerprinting

PCR Method

The invention of PCR technology advanced the DNA fingerprinting technology to a huge extent. Any specific region of DNA can be amplified by using oligonucleotide primers and thermostable DNA polymerase. Many different VNTR loci can be amplified and run on a gel; and in this way, samples can be distinguished. However, they were not as discriminating as RFLP. It was also difficult to determine a DNA profile for mixed samples, such as a vaginal swab from a sexually assaulted victim.

Amplified Fragment Length Polymorphisms

During 1990s, another technique was put forward for DNA fingerprinting or DNA profiling. It was called amplified fragment length polymorphism (AFLP). PCR was used to amplify samples, and therefore it was faster than RFLP. It relied on VNTR polymorphisms to distinguish various alleles, which were separated on a polyacrylamide gel using an allelic ladder and bands were visualized by silver staining the gel. A popular locus for fingerprinting was the D1S80 locus. As with all PCR-based methods, highly degraded DNA or very small amounts of DNA may cause allelic dropout (causing a mistake in thinking a heterozygote to be a homozygote) or other stochastic effects.

STR Analysis

In recent times, VNTR analysis has largely been replaced by STR analysis and is currently the standard approach to forensic DNA profiling. STR regions consist of 2–4 base-pair repeats, which are repeated 5–15 times. It covers a shorter region of DNA and is, therefore, very useful for forensic determination, as in many cases, only small DNA can be recovered from the crime scene and in some cases, even degraded DNA is obtained. STR analysis is faster, less labour intensive, and can be easily automated. Very little DNA is required for this analysis.

In STR analysis, PCR is used to amplify DNA in the region where the STR is located. These PCR products are then electrophoresed in a polyacrylamide gel (polyacrylamide gel is used as these products are of smaller size and cannot be run on an agarose gel that separates larger sized products). STRs are visualized by adding a DNA intercalator such as ethidium bromide into the gel, which intercalates into the DNA and emits fluorescence when excited by ultraviolet light. The migration of the PCR products is then compared to control DNA molecules that have a known size. If run together, the size of the unknown STR can be estimated. STR typing has been largely improved with the use of computer software with laser-controlled equipment.

Different STR-based DNA-profiling systems have been developed. Among these, the combined DNA index systems (CODIS) containing 13 STR loci is widely used. The 13 loci in CODIS are independently assorted. This means that if someone has the DNA type ABC, the probability of having the type ABC is equal to the probability of having A multiplied by the probability of having B multiplied by the probability of having C. With 13 loci, the probability of two random individuals having the same DNA profile is negligible.

STR analysis, however, is not without its drawbacks as well. If very little DNA is recovered from a crime scene and is degraded, not all regions in the genome will amplify, or there will be discriminatory amplification of DNA in only one chromosomal STR region rather than both. Additionally, there may be substances in the sample that inhibit the PCR reaction. However, the automation used for STR typing can lead to considerable reproducibility and quality control.

APPLICATIONS

The applications of DNA fingerprinting are many. It is very useful in identification of family relationship such as in the determination of paternity, in forensic crime analysis, in population genetics to analyse variation within population, and also in the diagnosis of inherited disorders.

Family Relationship Test

DNA analysis is widely applied to determine genetic family relationships such as paternity, maternity, and other kinships. During conception, a zygote contains a complete set of DNA molecules coming from both parents. During development, the DNA does not change and this allows the testing of relationship by DNA fingerprinting. The testing can be done with any type of sample, including skin cells, buccal swabs, blood, semen, or other tissue samples.

In a family relationship test, the digested DNAs from mother, father, and the putative babies are processed side by side. The pattern of bands found in the babies is compared with those found in the mother. All the bands that match in position and relative intensity could be maternal in origin. All the remaining bands in the DNA fingerprint must have come from the father. This is illustrated in Plate 1. Among the four babies tested, only two match the fingerprinting pattern of the father and mother.

Identification in Forensic Examinations

This is widely used in court cases for solving murder disputes. For example, specimens are collected from the crime scene and the DNA is extracted and matched with DNA samples taken from the body of the victim. The presence of foreign DNA can be predicted, and in some cases, if database is available, the suspect can be identified.

Medical Application

The extremely high individual specificity of DNA fingerprinting techniques is being used for monitoring bone marrow transplantation. VNTR probes are also used for monitoring in certain types of cancers.

Cell Line Authentication

Many of the experiments in molecular biology involve the use of specific cell lines. For accurate experiments, these lines should be specific and free of contamination from other cell lines. Any experiment such as the investigation of signalling pathways in neoplastic changes or treatment with some drug might give wrong conclusions. Cell line contamination is very common, and DNA fingerprinting can be used to test the authenticity of a cell line.

SUMMARY

- DNA fingerprinting detects polymorphisms in repeat sequence in a genome.
- Each individual has a unique pattern of these repeats.
- DNA is digested with restriction enzymes and hybridized to a suitably designed probe.
- DNA fingerprinting has applications in forensics and in personal identification.

REVISION QUESTIONS

1. What is DNA fingerprinting? Which types of markers are used for DNA fingerprinting?
2. What do VNTR and STR stand for? How are these used in genetic analysis?
3. How are VNTR patterns generated? Describe the procedure.
4. What is the principal difference between RFLP and VNTR fingerprinting?
5. Explain why VNTRs are never dominant as genetic markers.
6. Can DNA fingerprinting be used for species other than humans? Explain.
7. Why is the DNA sample first amplified by PCR in DNA fingerprinting?

12

RNAi and siRNA Technology

OBJECTIVES

After reading this chapter, the student will be able to:
- Learn about joining DNA fragments with cohesive ends
- Learn about the discovery of RNAi
- Understand the mechanism of RNAi
- Describe siRNA technology
- Discuss the synthesis of siRNA
- Analyse the applications of RNAi–siRNA technology

INTRODUCTION

RNA interference (RNAi) is a novel gene regulatory mechanism mediated by double-stranded RNA (dsRNA) in living cells to limit the transcript level of a gene by activating a sequence-specific RNA degradation process at the post-transcriptional level. It is often called post-transcriptional gene silencing, or PTGS. Two types of small RNA molecules—microRNA and small interfering RNA—are central to RNAi.

RNAi has an important role in defending cells from parasitic genes, viruses, and transposons and is also important in directing development and gene expression in general. The selective effect of RNAi on gene expression has made it a very valuable tool in molecular and cell biology research. Introduction of a synthetic dsRNA into cell culture and living organisms can induce suppression of the gene of interest. RNAi technology can also be a valuable tool in understanding the pathways involved in a cellular process, as the specific genes can be turned off to see its effect. Prokaryotes do not have RNAi systems homologous to eukaryotic ones, but some of the key proteins of eukaryotic RNAi can be identified among prokaryotic proteins involved in other functions. In mammals, RNAi system is more complicated, and small interfering RNA (siRNA), instead of long dsRNA, is involved in gene silencing.

DISCOVERY OF RNAi

In 1990, Rich Jorgensen and his colleagues introduced a pigment-producing gene under the control of a powerful promoter in the plant *Petunia* in order to deepen the purple colour of the flower. Instead of expressing deep purple colour, many of the flowers showed variegated or even white colours. Thus introduction of transgenes homologous to endogenous genes resulted in the suppression of both genes. The phenomenon was called co-suppression.

In 1995, Su Guo and Ken Kemphues used either antisense or sense RNAs to shut down the expression of *par*-1 in the nematode *Caenorhabditis elegans* in order to assess the function of the gene. Both antisense and sense RNAs were equally effective at silencing homologous target genes.

In 1998, Andrew Fire and Craig Mello injected dsRNA (a mixture of sense and antisense strands) into *C. elegans*. This injection resulted in the efficient silencing of expression of a gene that was homologous to the dsRNA. Injection of dsRNA retained the silencing activity till the next generation of the worm. Fire and Mello described this new technology based on the silencing of specific genes by dsRNA as RNAi. Shortly thereafter, RNAi was regarded as a breakthrough research tool that can be used to study disease-related pathways and can also be used in biotherapeutics. In 2004, it was used in the treatment of age-related muscular degeneration (AMD). In 2006, Fire and Mello shared the Nobel Prize in physiology and medicine for their work on RNAi in the nematode worm *C. elegans*, which was published in *Nature* in 1998.

MECHANISM OF RNAi

Using a combination of genetic studies and biochemical approaches, a cellular machinery mediating gene silencing has been identified in various organisms. A combination of results obtained from several in vivo and in vitro experiments suggest a two-step mechanistic model for RNAi post-transcriptional gene silencing (Plate 1).

The first step, or the initiation step, involves the digestion of the input dsRNA into a 20–23 nucleotide-long siRNA. The input dsRNA may result from the infection of a virus with an RNA genome or may be introduced directly in the laboratory. The dsRNA is directly imported into the cytoplasm. It is then cleaved into short fragments of siRNA by an enzyme Dicer. Dicer belongs to RNase III family of enzymes, which are dsRNA-specific ribonucleases. Dicer contains domains for dsRNA binding, RNA unwinding, and ribonuclease activity and drives the cleavage of dsRNA in an ATP-dependent manner. It makes staggered cuts in both the strands of dsRNA, leaving a 3' overhang of two nucleotides.

The initiating dsRNA may also come from endogenous sources, that is, from inside the cell. These are pre-microRNAs (miRNAs) expressed from RNA coding genes in the genome. The primary transcripts of these genes are processed to form the characteristic stem-loop structure of pre-miRNA in the nucleus and are then exported into the cytoplasm where these are cleaved by Dicer. Thereafter, these two pathways converge. The miRNAs also are able to silence gene expression by RNA cleavage, that is, behave as siRNAs. A current hypothesis is that a translational inhibition occurs if the target of miRNAs is imperfectly complementary to the target, whereas siRNA cleavage silences gene expression in cases of perfect complementarity (refer to Box 1 for more information on miRNAs). The distinctive characteristics of siRNAs and miRNAs are presented in Table 1.

In the second step, which is termed the effector step, these siRNAs join a multinuclease complex called RNA-induced silencing complex, or RISC. A precursor RISC is converted into a shorter complex on being activated by adenosine triphosphate (ATP). This activated complex cleaves the substrate. Under this activated condition of RISC, a helicase present in the complex might activate RISC by unwinding the siRNAs, which then become exposed. The antisense siRNAs then pair with the cognate mRNAs, and the complex cuts this mRNA approximately in the middle of the duplex region. The target cleavage site has been mapped to 11 or 12 nucleotides downstream of the 5'-end of the guide siRNA. A conformational rearrangement or a change in the composition of siRNA–RISC complex is postulated to occur ahead of the cleavage of target mRNA. Finally, the cleavage of the mRNA prevents translation. Following cleavage, the RISC complex disassembles and is ready to load another siRNA for the cleavage of additional mRNA.

Although the conversion of long dsRNA into many small siRNAs results in some degree of amplification, it is not sufficient to bring about such continuous mRNA degradation. Biochemical and genetic evidences suggest that RNA-dependent RNA polymerase (RdRP) plays a critical role in amplifying RNAi effects. A model of amplification suggests that RdRP uses the guide strand of an siRNA as a primer for the target mRNA, generating a dsRNA substrate for Dicer and thus generates more siRNAs.

Table 1 Distinctive characteristics of miRNA and siRNA

miRNA	siRNA
Double-stranded RNA of 21–23 base pairs	Double-stranded RNA of ~21 base pairs with 3'-dinucleotide overhangs
Endogenous non-protein coding RNA molecule	Generated by either degradation of exogenous (for example, viral) dsRNAs or transcribed from transposable elements in the genome
Encoded by distinctive miRNA genes	There are no dedicated genes
Shows imperfect complementarity to their targets	Shows perfect complementarity to their targets, even one base pair mismatch reduces their activity by 90%

Box 1   Biogenesis of miRNAs

MicroRNAs, or miRNAs, are double-stranded RNAs that are ~22 nucleotides long on an average and are derived from hairpin-structured precursors. These are involved in the regulation of gene expression in eukaryotes usually by post-transcriptional silencing. MicroRNAs are found in almost all metazoans such as worms, flies, zebra fish, and mammals. Genes for miRNA are located in all chromosomes and are found in introns as well as exons. These miRNAs are transcribed by RNA polymerase II as precursor molecules (pri-miRNA) with 5′ m7G capping structures and 3′ poly(A) tail, which are subsequently cleaved by the Drosha-Pasha complex to produce a stem-loop-structured precursor miRNA called pre-miRNA. Drosha is an RNase III family protein and cleaves the precursor transcript at the base of a stem structure with the help of Pasha, which is a double-stranded RNA-binding protein. This processing occurs in the nucleus. The pre-miRNAs are then recognized by the nuclear export factor, Exportin-5 (Exp5), which partners with the Ran GTP-binding protein to form a nuclear transport complex that delivers pre-miRNA through nuclear pores to the cytoplasm. It is then chopped into a 22-nucleotide-long duplex mature miRNA by Dicer, which is a highly conserved cytoplasmic RNase III. In the next stage, miRNAs are loaded onto the RISC complex like siRNAs for gene silencing.

The mechanism of gene silencing by miRNA, like siRNAs, involves the interaction of miRNAs with the target mRNAs. The binding occurs through imperfect base pairing between the miRNA guide strand and its target mRNA. Due to the absence of perfect complementarity, and hence the presence of mismatches, miRNA guide strands usually form bulge structures. The sequence specificity is generally determined by 2–8 base pairs of miRNAs, usually called the seed region. The plant and animal miRNAs differ in their specificity. The short seed region is required for the functioning of RISC complex. Since miRNAs can mediate gene silencing even if there is imperfect pairing, this raises the potential for a single miRNA to target multiple mRNAs. However, better understanding of the miRNA function in future would guide the development of efficient biochemical and genetic approaches to determine miRNA targets in vivo.

RNAi is a physiological response, which has been utilized for experimental control of gene expression. The RNAi technology finds its application in *C. elegans*, *Drosophila*, plants, and protists. A long (>500 base pairs) dsRNA is commonly used for silencing. For the introduction of dsRNA, several methods are used. These are microinjection, feeding, soaking, or transfection. Alternatively, transgenic organisms can be made, which express dsRNA from transgenes. A number of studies have been reported demonstrating RNAi in varied organisms; a selective list is presented in Table 2.

The advantages of RNAi technology are as follows.

❖ A loss-of-function phenotype can be readily linked to a specific gene.

❖ Knocking down a gene by this technology does not knock down the gene permanently so that turning on and off of a gene can be done many times in an animal's life.

Table 2	Eukaryotic organisms exhibiting	
Kingdom	*Species*	*Delivery method*
Protozoans	*Trypanosoma brucei*	Transfection
	Plasmodium falciparum	Electroporation and soaking
	Toxoplasma gondii	Transfection
Invertebrates	*Caenorhabditis elegans*	Transfection, feeding bacteria carrying dsRNA, soaking
	Caenorhabditis briggsae	Injection
	Brugia malayi (filarial worm)	Soaking
	Drosophila melanogaster	Injection soaking and transfection
Vertebrates	Zebra fish	Microinjection
	Xenopus laevis	Injection
	Mice	Injection
	Human cell lines	Transfection
Plants	Monocots/dicots	Particle bombardment with siRNA/transgenics
Fungi	*Neurospora crassa*	Transfection
	Dictyostelium discoieum	Transgene
	Schizosaccharomyces pombe	Transgene

siRNA TECHNOLOGY

RNAi, the dsRNA-induced gene silencing, has been observed in a variety of organisms, including plants, protozoa, insects, and nematodes. However, in the mammalian system, the response to dsRNA was found to be different and more complicated. First of all, when a dsRNA encounters a mammalian cell, the latter mistakes the dsRNA as some viral product and develops immune response involving the release of interferon, which ultimately kills the cell. Furthermore, the transfection of dsRNA causes non-specific suppression of gene expression as opposed to specific suppression observed for others. Interestingly, it was found later that if siRNAs (<50 nucleotides) are introduced into mammalian cells, these avoid the activation of the antiviral response mechanism and a gene-specific silencing process can be initiated. Thus it was found that in mammalian cells, RNAi could be prompted through the use of shorter pieces of RNA known as siRNAs to silence specific genes without activating the interferon response.

The effectiveness of siRNA varies; the most potent siRNA is able to reduce about 90% of the protein expression in cultured cells. Here also like worms, the most effective siRNA turned out to be a 21-nucleotide dsRNA with two nucleotides overhanging at the 3'-end. Sequence specificity of siRNA is very stringent; a single base pair mismatch between the siRNA and the target mRNA decreases the silencing dramatically. This property of siRNA is very useful in allele-specific silencing in case one allele contains an SNP, or single nucleotide polymorphism. However, not all siRNAs with this specification will turn out to be effective. The design of siRNA is, therefore, an important procedure to be considered for application of siRNA technology.

The siRNA technology for gene silencing is rapidly evolving and finding wide use in deciphering the cellular pathways as well as in therapeutics. Gene silencing using siRNA involves the following steps.

❖ Design of siRNA
❖ Synthesis of siRNA
❖ Delivery of siRNA
❖ Monitoring knock-down of the gene

Design of siRNA

The reduction in the mRNA level depends on the target site at which siRNA binds. The first step in designing an siRNA is to choose the siRNA target site, meaning the region of mRNA to be targeted. Design is done with bioinformatics, keeping in mind several experimental observations.

There are several criteria, as listed below.

❖ The 21-nucleotide sequences in the target mRNA beginning with the dinucleotide AA will be a potential site. This strategy emerges from the observation of Elbashir and his colleagues that siRNAs with UU overhangs are most effective.

❖ Stretches of nucleotide repeats should be avoided; presence of three or more stretches of G or C initiates intramolecular secondary structures preventing siRNA hybridization. Presence of stretches of A to T reduces the specificity of target sequence.

❖ GC content should be less than 50%.

❖ Choice of 5'- and 3'-UTR sequences is avoided, although in some cases, siRNAs targeting UTRs have successfully induced gene inhibition.

❖ The siRNA sequence should not match other gene sequence so that the silencing may become non-specific. This can be checked through the Basic Local Alignment Search Tool, or BLAST, available at NCBI.[1] Sequences with more than 15 contiguous base pairs of homology to other genes in the NCBI database are usually eliminated.

Generally, two to four siRNAs are designed, and a negative control is used. The negative control is a sequence of the same length as that of the siRNA—same sequence composition (same number of As, Ts, Gs, and Cs) but with a different sequence specificity. (An siRNA design tool is also publicly available at the Whitehead Institute of Biomedical Research, MIT).[2]

Synthesis of siRNA

There are several methods for preparing siRNA, such as chemical synthesis, in vitro transcription, siRNA expression vectors, and polymerase chain reaction (PCR) expression cassettes. Custom-made siRNAs are available from various organizations.

---

[1] Details available at <www.ncbi.nlm.nih.gov/BLAST>
[2] Details available at <www.jura.wi.mit.edu/bioc/siRNAext/>

To synthesize siRNA by in vitro transcription, at first, two DNA oligonucleotides (with eight bases complementary to T7 promoter primer and containing sense and antisense sequences) are annealed to T7 promoter primer and filled in with Klenow. These are then transcribed with T7 RNA polymerase when both sense and antisense RNA products are obtained. The RNA strands are then hybridized to create siRNAs.

Many siRNA expression vectors are available. Here a designed siRNA target sequence is used to synthesize the corresponding DNA insert. The DNA insert is then cloned to an siRNA expression vector containing RNA polymerase III promoter, which can transcribe small nuclear RNAs efficiently and is then directly transfected into the cell. These plasmids will produce the desired siRNA directly inside the cell, as shown in Figure 1.

Delivery of siRNA

The delivery of siRNA to cell lines can be done by transfection. Two methods are available: chemical method and electroporation. Many variables affect the efficiency of transfection. The transfection optimized for DNA or oligonucleotides normally does not work well with siRNAs. So the procedure for siRNAs has to be optimized. A polyamine mixture or a mixture of cationic and neutral lipids is used for the delivery of siRNA in cells. In electroporation, an electric pulse is applied, which results in transient pore formation in the cell membrane through which siRNAs pass. The pores then reseal and the cells recover.

Monitoring Knock-down by siRNA

This is done by checking the expression of the targeted gene by real-time PCR or other methods using some housekeeping gene as a control.

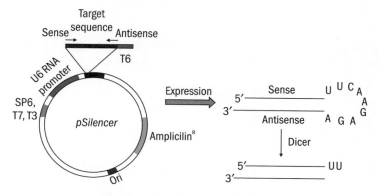

Figure 1　siRNA expression vector

APPLICATIONS OF RNAi–siRNA TECHNOLOGY

The advent of RNAi technology has increased the possibility of therapeutic interventions in those cases that were previously thought impossible. Traditional drug targets, be it proteins, enzymes, or receptors, belong to post-translational targeting, whereas siRNAs or RNAi technology is post-transcriptional, and if designed properly, siRNAs can be very specific. Various medical applications are building excitements, and some potential applications and the targets are listed in Table 3.

SUMMARY

- RNA interference is a mechanism operative inside a cell for regulating gene expression.
- Two types of small double-stranded RNA molecules, namely, miRNA and siRNA, are involved in RNA interference.
- RNAi is achieved at post-transcriptional level when siRNA binds to mRNA, resulting in its cleavage.
- Organisms vary in their ability to take up foreign dsRNA and use it in the RNAi pathway.
- In most mammalian cells, shorter RNAs are used as these cells show immune response against long double-stranded RNA molecules.
- Specialized laboratory techniques like siRNA expression vectors are used for siRNA delivery in mammalian cells.
- RNAi or siRNA technology has found widespread application as laboratory research technique for altering the amount of specific proteins inside cells.
- There is also active study on the medical applications of RNAi.

Table 3 Examples of diseases and silencing of respective targets by RNAi–siRNA technology

Disease	Protein targeted
Chronic myeloid leukaemia	Bcr/Abl
Neuroblastoma	N-Myc
Carcinomas	Ras
Fragile X syndrome	dFMR1
Hepatitis	Fas
Acute liver failure	Caspase 8
Malaria	Falcipain-1,2
HIV	p24 Gag, CCR5
Rotavirus	VP4
Influenza	Nucleocapsid (NP) RNA transcriptase
Drug resistance in pancreatic and gastric carcinoma	MDR1

REVISION QUESTIONS

1. What role did *Caenorhabditis elegans* and *Petunia* play in the discovery of RNAi?
2. How does RNAi work? Why did cells evolve this mechanism?
3. What is the function of the RISC–RNA complex?
4. What are the possible sources of double-stranded RNA in cells? Why did cells evolve enzymes that recognize and destroy double-stranded RNA molecules?
5. Why cannot gene regulation in mammalian cells be done by long double-stranded RNAs? What technique is used instead?
6. How can RNAi be used to treat a disease? What is the evolutionary function of RNAi?

13

Molecular Biology Methods

OBJECTIVES

After reading this chapter, the student will be able to:
- Describe joining DNA fragments with cohesive ends
- Understand cloning blunt-ended DNA fragments
- Learn the use of oligonucleotide primers
- Explain the use of adapter molecules
- Describe the cloning of PCR products (TA cloning)

INTRODUCTION

This chapter covers the detailed aspects of the experimental strategies adopted in molecular biology. The experiments are done by molecular biologists to elucidate the structure and function of genes and genomes, which, in turn, help us to understand life processes at the molecular level. This chapter will focus on the major experimental techniques of molecular biology.

In the earlier chapters on recombinant DNA technology, the basic cloning procedure has been described. Essentially, any cloning procedure involves the following.

❖ Generation of DNA fragment, which may be a single gene, a PCR product or may have a mixture of fragments as in case of DNA library construction.

❖ Insertion of the DNA fragment into the cloning vector, which involves design of several strategies depending on the nature of the vector and insert.

❖ Introduction of the vector into the recipient cell; for plasmid vectors, this is usually done by the transformation process.

❖ Selection of the recipient cells that have acquired the recombinant plasmid. The strategies for the above are illustrated schematically in Figure 1.

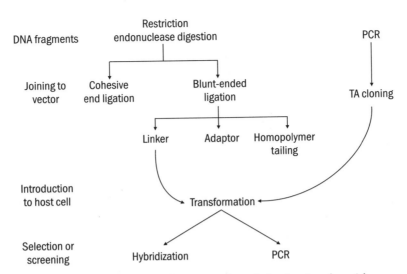

Figure 1 Schematic for strategies of cloning in plasmids

In this chapter, we will discuss the various strategies involved in the insertion of the DNA fragment into the plasmid.

This chapter will provide the detailed aspects of the experimental strategies adopted. The experiments elucidate the structure and function of genes and genomes, which help in understanding life processes at the molecular level.

JOINING DNA FRAGMENTS WITH COHESIVE ENDS

The easiest DNA fragments to clone are those that have cohesive ends, or in other words, protruding 5'- or 3'-termini. These single-stranded termini are created by the digestion of the vector and the target DNA with restriction enzymes that cleave DNA asymmetrically within the recognition sequence. The generation of compatible sticky-ended DNA fragments makes the vector and insert associate with each other through complementary base pairing, but they will not form a continuous sugar–phosphate DNA backbone. DNA ligase is required for further joining of these ends.

DNA ligase has the capability of sealing single-stranded nicks between adjacent nucleotides in a duplex DNA chain. This enzyme is produced by *Escherichia coli* as well as by the bacteriophage T4. Although the reactions catalysed by the enzymes of *E. coli* and T4-infected *E. coli* are very similar, they differ in their cofactor requirements as well as in specificity. The cofactor required by T4 enzyme is ATP, while the *E. coli* enzyme requires NAD^+. In each case, an enzyme–AMP complex is formed, which binds to the nick and makes a covalent bond between the exposed 5'-phosphate and 3'-OH group, sealing the nick.

When the vector and DNA fragments with cohesive ends associate, the joint has nicks that are a few base pairs apart in opposite strands. DNA ligase is added in vitro, which then repairs these nicks to form an intact duplex (Plate 1). The optimum temperature for ligation of nicked DNA is 37°C, but at this temperature, the hydrogen bond between the sticky ends is unstable. *Eco*RI-generated termini associate through only four AT base pairs, and these are not sufficient to resist thermal disruption at such a high temperature. The optimum temperature for ligating the cohesive termini is, therefore, a compromise between the rate of enzyme action and association of the termini and has been found experimentally to be in the range of 4–15°C.

Further, formation of a circular recombinant plasmid DNA molecule capable of transforming to the host occurs in a two-step process, namely, an intermolecular reaction between the linear plasmid and the incoming DNA creating a linear hybrid with cohesive termini, followed by an intramolecular reaction in which the cohesive ends of the linear hybrid are joined together to form a circular molecule. The ligation reaction might yield a number of potential products like desirable monomeric circular recombinant plasmid, and undesirable linear and circular homopolymers and heteropolymers of various sizes and compositions. The ligation reaction has to be monitored, for example, to favour the formation of circular monomeric recombinants. The number of recombinants can be increased by using high DNA concentration in a ligation reaction. However, at low DNA concentration, the probability of the formation of monomeric circular recombinant molecules is relatively favoured because of the reduced frequency of intermolecular reactions.

To prevent recircularization of plasmid DNA or plasmid dimer formation, the linearized plasmid vector DNA is treated with alkaline phosphatase to remove the phosphates from the free 5′-ends of the cut DNA. The strategy is illustrated in Plate 2. The ligation of a compatible insert into this vector will result in the ligation of only the 5′-ends of the insert with the vector. The 3′-end of the insert (a hydroxyl group) and the 5′-end of the vector (also a hydroxyl group) will be unable to ligate. In this process, one nick at each joint remains unligated, but after transformation into the host bacteria, the broken DNA strands will be repaired using the bacterial DNA repair systems, forming the complete vector plus insert plasmid.

CLONING BLUNT-ENDED DNA FRAGMENTS

Some restriction enzymes (for example, *Eco*RV, *Sma*I) leave blunt ends in the DNA fragment. Fragments of foreign DNA with blunt ends can be cloned into a linearized plasmid vector bearing blunt ends. The *E. coli* DNA ligase will not catalyse blunt ligation except under special reaction conditions. Only T4 DNA ligase is able to join blunt-ended DNA molecules. Ligation of blunt-ended fragments is a comparatively inefficient process but can be optimized under certain conditions like low concentration of ATP, absence of inhibitors like polyamine or spermidine, high concentration of ligase, and high concentration of blunt-ended fragments. For efficient blunt-ended ligation, several other strategies are used, for example, use of linkers and adapters.

Using Oligonucleotide Linkers

Linkers are small (8–16 nucleotides) self-complementary pieces of synthetic DNA that anneal to form blunt-ended, double-stranded molecule containing sites for one or more restriction endonucleases. Linkers are at first ligated to both ends of the foreign DNA (which are blunt ended) to be cloned. Digestion with restriction endonuclease will then produce sticky-ended fragments, which can further be introduced into a vector digested with the same restriction enzyme. A large variety of linkers having varying restriction sites are available from commercial sources. Joining of DNA molecules with one such linker is illustrated in Figure 2.

Using Adapter Molecules

Sometimes, it may happen that the restriction enzyme used to generate the cohesive ends in the linker may also cut the foreign DNA at internal sites. In such a case, two or more sub-fragments of the foreign DNA will be cloned. Replacement of one linker

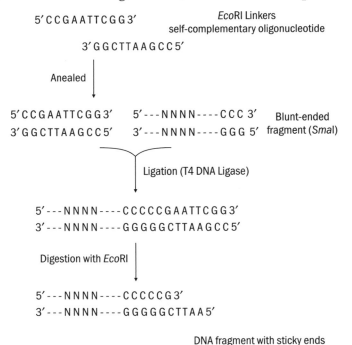

Figure 2 Linkers

by another may be done, but this would not be the right choice in case of cloning large DNA fragments, as these are expected to contain sites for a number of restriction enzymes. A solution to this problem could be provided by the use of adapters. Adapter molecules are short double-stranded synthetic oligonucleotides with single-stranded tails at one or both ends and also carry an internal restriction endonuclease recognition site. By adding adapters to the ends of a DNA, sequences that are blunt can be converted into cohesive ends. For example, a blunt-ended foreign DNA may be ligated to an adapter molecule containing one blunt end bearing a 5'-phosphate group and a *Bam*HI cohesive end, which is not phosphorylated. The foreign DNA plus added adapters are then phosphorylated at the 5'-termini and ligated into the *Bam*HI site of the vector.

Alternatively, if a foreign DNA to be cloned contains an internal *Eco*RI site, it cannot be digested by *Eco*RI for cloning into an *Eco*RI cut vector. In such a case, the foreign DNA could be digested with another enzyme say *Bam*HI and using adapter, *Eco*RI cohesive end can be generated. Such an example is illustrated in Figure 3. The sequences 5'-GATCCCCGGG-3' and 5'-AATTCCCGGG-3', when used as adapters, will produce a *Bam*HI cohesive end at one end and an *Eco*RI sticky end at the other side of the insert without the need for cutting with the specific restriction enzymes. Ligation of this adapter to a restriction fragment generated by *Bam*HI digestion will produce a DNA fragment with *Eco*RI ends that can now be ligated with an *Eco*RI digested vector. A wide range of adapters is available commercially. Note that the only difference between an adapter and a linker is that the former has cohesive ends and the latter has blunt ends.

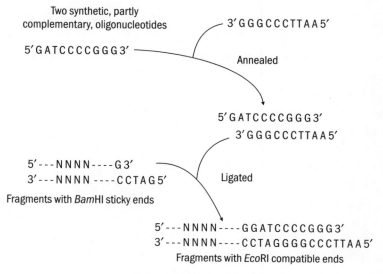

Figure 3 Adapters

Homopolymer Tailing

Homopolymer tailing is a technique in which a homopolymeric tract of a nucleotide is attached to the end of a DNA fragment by terminal transferase (or terminal nucleotide transferase). The enzyme terminal transferase, when supplied with deoxynucleotide triphosphates (for example, dTTP), adds nucleotides to the 3'-OH end of a DNA molecule. Unlike DNA polymerase, terminal transferase does not require a template strand, so this enzyme will add T residues at each 3'-end and hence it is given the term "homopolymer tailing". This is an alternative way to add sticky ends to DNA molecules and is useful for cloning. For example, if oligo(dA) sequences are added to the 3'-ends of the foreign DNA fragment and the vector DNA is tailed by oligo(dT), the two types of molecules can anneal to form mixed dimeric circles by DNA ligase. DNA with exposed 3'-OH groups may be generated by pretreatment with phage lambda exonuclease. The main advantage of tailing is that it could provide longer regions of sticky ends compared to restriction digested sticky ends and increase the specificity of ligation.

CLONING OF PCR PRODUCTS (TA CLONING)

The amplification procedure used in polymerase chain reaction (PCR) is expected to yield blunt-ended products. The blunt-ended cloning procedure may then be used for cloning. However, in practice, the efficiency of blunt-ended cloning is very low. This has been traced to the fact that the Taq DNA polymerase adds a non-specific adenosine residue to the 3'-end of double-stranded DNA. As a result, the product is not truly blunt ended, and blunt-ended and cloning is not successful. The uses of linkers are not successful.

An alternative strategy is to use a modified primer that contains an additional restriction enzyme recognition site at its 5'-end. The resulting product can then be cut with the restriction enzyme and cloned to a plasmid cut with the same restriction enzyme. In this way, a PCR product can be efficiently cloned.

Another direct cloning method exploits the fact that the Taq DNA polymerase adds a non-specific adenosine residue to the 3'-end of a product. A complementary vector is engineered to contain a 3'-overhang of T residues. This produces short but sticky ends, which will anneal, allowing quite efficient ligation. The basic strategy for cloning PCR products with TA cloning vectors is presented in Figure 4.

A further improvement to PCR cloning is the use of vaccinia virus DNA topoisomerase I (TOPO). In vivo topoisomerases are involved in the supercoiling/relaxation of circular DNA cleaving DNA at specific sites, leaving a sticky end. The energy released by the breaking of phosphodiester bond is stored in a covalent bond between the enzyme and one of the cleaved strands. The enzyme trapped in the sticky end will then rapidly release its stored energy to form a new phosphodiester bond as soon as the sticky end encounters its complementary partner. Thus TOPO has both endonuclease and ligase

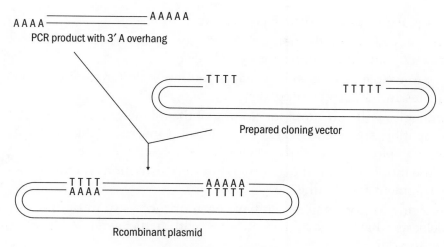

Figure 4 TA Cloning

captivity. Commercially available TOPO has sticky-end overhang and can be used for cloning PCR products efficiently.

SUMMARY

- A cloning procedure requires insertion of the DNA fragment into the suitable cloning vector and is done with the use of strategies depending on the nature of the vector and insert.

- Digestion of the vector and insert DNA with restriction enzymes that cleave DNA asymmetrically within the recognition sequence generates compatible sticky-ended DNA fragments of vector and insert. The vector and insert associate with each other through complementary base pairing, which are then sealed by DNA ligase.

- Some restriction enzymes such as *Eco*RV, *Sma*I produce blunt-ended double-stranded DNA fragments. For ligation, these blunt-ended DNA molecules are converted to cohesive ends by the use of linkers or adapters.

- Cloning of PCR products is done by exploiting the fact that the Taq DNA polymerase adds a non-specific adenosine residue to the 3'-end of a product. A complementary vector is engineered to contain a 3'-overhang of T residues. This produces short but sticky ends, which will anneal allowing quite efficient ligation.

REVISION QUESTIONS

1. Why is it useful to take plasmids carrying antibiotic resistance genes as vectors? What happens if the cloned DNA sequence is inserted into the middle of an antibiotic resistance gene?

2. How is a plasmid designed to serve as a cloning vector?

3. Why would you normally not use restriction endonucleases with four or eight nucleotide cut sites for cloning into plasmids?

4. What is a multiple cloning site? What is the rationale for the inclusion of multiple cloning sites in modern cloning vectors?

5. What strategies will you adopt for cloning of the following: (1) both vector and insert have compatible sticky ends, (2) both vector and insert are digested with *Sma*I, and (3) insert has blunt end and the vector has sticky end?

6. What is the rationale behind using two different restriction endonucleases to prepare a DNA fragment for cloning?

7. You are trying to clone a gene but have been unsuccessful in finding any restriction endonucleases that do not have cut sites within the gene. What alternatives are available to you?

14

Features of Commonly Used Vectors

OBJECTIVES

After reading this chapter, the student will be able to:
- Explain plasmids as cloning vectors
- Explain the desirable properties of plasmid cloning vectors
- Discuss plasmid vectors pBR322 and pUC18
- Describe bacteriophage lambda vector and M13 vector
- Understand cosmid vectors
- Explain artificial chromosome vectors
- Describe bacterial artificial chromosomes
- Discuss yeast artificial chromosomes
- Explain mammalian artificial chromosomes

INTRODUCTION

Cloning vectors are molecules that act as carriers of foreign DNA in genetic engineering. The commonly recognized features of cloning vectors are as follows.

❖ They can replicate independently inside their host with or without foreign DNA.

❖ They contain a number of restriction enzyme sites, which are present only once.

❖ They carry selectable marker(s) (for example, ampicillin resistance gene or gene for enzymes missing in the host cell) to distinguish host cells carrying vectors from those that do not carry these vectors.

❖ They could be relatively recovered in an easy way from the host cells.

There are several possible choices of vectors like plasmids, bacteriophages, cosmids, bacterial artificial chromosomes (BACs), yeast artificial chromosomes (YACs), and mammalian artificial chromosomes (MACs) (Table 1).

Table 1 Principal features and applications of different cloning vectors

Vector	Maximum insert size	Examples	Major applications
Plasmid	10–20 kb	pBR322, pUC18	General DNA manipulation, subcloning and downstream manipulation, cDNA cloning and expression assays
Phage λ	10–20 kb	λgt11, λZAP, EMBL4	Genomic DNA cloning, cDNA cloning and expression libraries
M13 phage	8–9 kb	M13mp18	In vitro mutagenesis, DNA sequencing
Cosmid	35–45 kb	pJB8, SuperCos-1	Genomic library construction
Bacterial artificial chromosome (BAC)	75–300 kb	pBAC108L, pBeloBAC11	Analysis of large genomes
Yeast artificial chromosome (YAC)	100–1000 kb (1 Mb)	pYAC4	Genome mapping, molecular cytogenetic studies, FISH banding, YAC transgenic mice
Mammalian artificial chromosome (MAC)	100 kb to > 1 Mb		Under development for use in animal biotechnology and human gene therapy

PLASMIDS AS CLONING VECTORS

Plasmids are widely used as vectors for cloning of small DNA fragments. A plasmid vector should have a DNA region called an origin of replication, or origin. This region allows the plasmid to replicate independently inside the host. If a foreign DNA is inserted into the plasmid, it replicates its own DNA along with the foreign DNA, thus generating many copies of the recombinant plasmid. For typical cloning experiments, the plasmid DNA must contain some restriction sites that cut the DNA only once so as to make the circular DNA linear. If it cuts more than once during ligation, some fragment of the plasmid DNA is lost.

DESIRABLE PROPERTIES OF PLASMID CLONING VECTORS

A bacterial plasmid vector should have some or all of the following properties to function as an ideal cloning vehicle.

❖ It should be small, that is, of low molecular weight. Small plasmids are easier to handle, replicate faster (high copy number), and require less energy for replication than large ones. Small plasmids are less susceptible to damage by shearing and are easier to purify than large ones. For small plasmids, there is a probability of getting more restriction sites that are unique.

❖ The DNA sequence is known. With the availability of complete DNA sequence, the complete utilization is possible.

❖ A high copy number plasmid will produce large quantities of foreign DNA, which will be useful for many applications.

❖ Unique restriction sites should be present for many enzymes.

❖ Selectable markers should be present, as frequently used plasmids contain ampicillin resistance gene or tetracycline resistance gene as selectable marker. The presence of any of these markers enables the bacterial cell to grow in the presence of ampicillin or tetracycline when it contains the plasmid. In the absence of plasmid, the bacterial cells should be sensitive to these antibiotics, thus enabling suitable detection of these plasmids.

❖ A second selectable gene should be present, which is disrupted upon insertion of foreign DNA. It helps in the identification of the foreign DNA. Suppose a plasmid contains two antibiotic resistance markers A and B. If the foreign DNA is inserted in such a way that gene B is disrupted but not gene A, the bacterial host containing recombinant DNA (plasmid vector with foreign DNA) will be sensitive to B but resistant to A. The bacterial host containing a vector only will be resistant to both A and B.

❖ Multiple cloning sites should be present inside the marker. If a large number of restriction sites are available internal to a marker (often called multiple cloning site), the vector is versatile and suited for cloning varied fragments.

❖ Presence of termination sequences.

PLASMID VECTORS pBR322 AND pUC18

Early cloning experiments were conducted by using naturally occurring plasmids like pSC101, ColE1. These plasmids had the advantages of being small and had a single site for common restriction enzymes; they suffered from the limitation of availability of genetic markers for screening. So later on, several superior plasmid cloning vectors were developed to suit the need of various applications. The popular plasmid vectors used for all purpose cloning are pBR322 and pUC series of vectors. The characteristics of these are widely used vectors, as described subsequently (Figure 1).

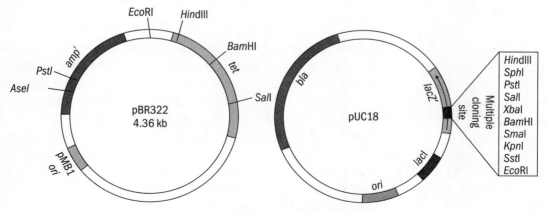

Figure 1   Structure of pBR322 and pUC18

pBR322

The plasmid pBR322 was the first artificial plasmid designed by Boliver and Rodriguez in the 1970s, and at one time, it was the most widely used plasmid vector. It was constructed by using both classical genetic techniques and recombinant DNA methodology. The vector contains two selectable markers: an ampicillin resistance gene taken from RSF2124 and a tetracycline resistance gene from pSC101 plasmid. It contains ColE1 origin of replication element from the plasmid pMB1, which is relaxed, enabling the plasmid to accumulate in high numbers in its host *Esherichia coli*. Its complete sequence is known to contain 4361 base pairs. Thus it can be characterized in terms of restriction sites and fragments generated thereafter. It contains single sites for over 40 restriction enzymes, of which 11 are in tetracycline resistance (*tet*^{r}) gene and 6 in ampicillin resistance (*amp*^{r}) gene. Cloning with any of the enzymes present in *tet*^{r} or *amp*^{r} gene will result in the insertional inactivation of either marker, while cloning in other unique sites will not permit easy selection of recombinants, as neither of the above genes will be inactivated.

pUC18

Another common plasmid artificially prepared and widely used in recombinant DNA technology is pUC18. This vector was constructed by using the region containing the origin of replication *oriV* of the plasmid ColE1. For the selection of those cells that have received the plasmid DNA, pUC18 contains an antibiotic resistance gene β-lactamase, which confers resistance to ampicillin. There is a second marker carried by pUC18, namely, a β-galactosidase gene. pUC18 contains a fragment of β-galactosidase gene coding for alpha-fragment, which can complement the host inactive β-galactosidase gene. Transformation of the host by pUC results in cells producing functional β-galactosidase. They are then detected by the chromogenic substrate X-gal, which gives a blue product when hydrolysed by β-galactosidase. When the host contains pUC vector, blue colonies will result.

In pUC18, a piece of synthetic DNA has been inserted, which contains recognition sites for a number of restriction enzymes. This multiple cloning site allows considerable flexibility in the choice of restriction enzymes. The enzymes are chosen in such a way that there is only one recognition site for a particular enzyme, so that digestion with any one may give a linear plasmid. Insertion of a DNA fragment results in the inactivation of the gene in which that site is found. The cloning site must, therefore, be chosen from a region of plasmid that is non-essential for the replication of plasmid, but at the same time, plasmid with foreign DNA can be distinguished from plasmid without DNA. pUC18 contains the multiple cloning site within the initial sequence of the β-galactosidase gene (but does not inactivate the β-galactosidase gene). If a DNA fragment is inserted into the multiple cloning site, it will (usually) prevent the production of β-galactosidase, either by interrupting transcription or by altering the reading frame. This is known as insertional inactivation. Genuine recombinants will, therefore, be white on a medium containing X-gal and can be distinguished from blue colonies containing the original pUC18.

PHAGE VECTORS

Bacteriophage Lambda Vector

Bacteriophage lambda (λ) is a temperate bacteriophage that has two alternative life cycles—a lytic cycle and a lysogenic cycle (Plate 1 of Chater 5). When reproducing in its lytic cycle, it undergoes general recombination as much as phage T4 does. The bacteriophage λ contains linear double-stranded DNA. However, unlike T4, λ DNA does not show terminal redundancy, nor are the phages circularly permuted. Unlike the DNA molecules in T4 phage, every phage λ DNA molecule has identical ends. However, the ends are single stranded and complementary in sequence so that they can pair, forming a circular molecule. The single-stranded ends are called cohesive ends to indicate their ability to undergo base pairing. The packaging of DNA in phage λ does not follow a headful mechanism like T4. Rather, the λ packaging process recognizes specific sequences that are cleaved to produce the cohesive ends.

The λ map is illustrated in Figure 3 of Chapter 5. The genes show extensive clustering by function. The left-half of the map entirely consists of genes encoding head and tail proteins, and within this region, the head genes and the tail genes themselves form subclusters. The right-half of the λ genome shows gene clusters for DNA replication, recombination, and lysis. The genes are clustered not only by function but also according to the time at which their products are synthesized. For example, the N gene acts early; genes *O* and *P* are active later; and genes *Q, S, R*, and the head-tail cluster are expressed last. The transcription patterns for mRNA synthesis are thus very simple and efficient. There are only two rightward transcripts, and all late genes except for *Q* are transcribed into the same mRNA.

Replication and Packaging of Lambda

When λ infects *E. coli*, the linear DNA is injected into the bacterial cell, where it is converted into a circular form due to the presence of *cos* sites. These *cos* sites contain sequences complementary to one another and form base pairs with one another, thus creating a circular molecule. The DNA ligase within the cell then rapidly seals the nicks in the circle to form a covalently closed circular DNA molecule.

The covalently closed DNA molecule initially replicates by the theta mode, which is essentially similar to the replication of other circular molecules, such as plasmids. At a later stage in infection, there is a switch from the theta mode to rolling circle replication, which yields a multiple length linear DNA molecule (Figure 2). Like T4 phage, tails and empty heads have been produced at this stage, but the packaging mechanism is different. With λ, the extent of the DNA to be packaged is determined by the position of the *cos* sites. A protein attached to the phage head recognizes a *cos* site in the multiple length DNA molecule and initiates the packaging of the DNA. This proceeds until the next *cos* site is reached, when the protein cuts the DNA at each *cos* site. These cuts are asymmetric, that is, the two strands are cut at positions that are not opposite to one another. There is a distance of 12 bases between the two cuts, leading to a sequence of 12 unpaired nucleotides at each end of the packaged linear DNA. The head is then sealed, the preformed tail is added, and the mature phage particles are released by lysis of the cell.

This contrasting packaging mechanism is of practical importance in the use of λ for gene cloning. Phage λ has packaging limits; the distance between two *cos* sites must be between 75% and 107% of the wild-type sequence (that is, between 37 kb and 52 kb; wild-type lambda DNA has 48.5 kb of DNA).

Wild-type λ DNA contains a number of target sites for most of the commonly used restriction endonucleases and is not itself suitable as a vector. Derivatives of the wild-type phage have, therefore, been constructed and are mainly of two types: insertional vectors in which foreign DNA can be inserted into a specific restriction

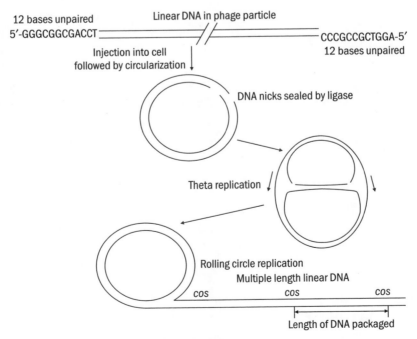

Figure 2   Replication of bacteriophage lambda DNA

enzyme recognition site and replacement vectors where foreign DNA replaces a piece of DNA (stuffer fragment) of the vector. The replacement vectors contain two target sites for restriction endonucleases. Cutting the DNA at these two sites will result in the production of three fragments: the left fragment, the right fragment, and a central portion, which is called the stuffer fragment. The stuffer does not contain any essential gene and can, therefore, be separated and discarded. The left and right fragments when joined are too small to allow viable phage particles to be formed. Such particles require an additional inserted piece of DNA for the formation of viable phage particles, which must be, in this case, at least 7 kb and not more than 22 kb. Therefore, not only is the cloning capacity increased, but such vectors are selective for reasonably large inserts. Vectors of this type are known as replacement vectors. Certain features of wild-type λ DNA posed limitations to its use as suitable vector. It contains a few unique restriction enzyme recognition sites for cloning of foreign DNA fragments, and the packaging of DNA into the λ phage is size limited. Efficient packaging occurs with DNA fragments that are 78% and 105% of the wild-type genome size. However, two important developments have enabled the use of λ as vector. First, the gene products required for recombination could be removed from the λ genome, and the remaining DNA is capable of providing all necessary functions for the lytic pathway to occur. Second, naturally occurring restriction enzyme recognition sites could be eliminated without the loss of gene function, permitting the development of unique sites for the insertion of foreign DNA. Lambda vectors are capable of carrying much larger foreign DNA inserts compared to plasmid vectors. Two basic types of λ vector have been developed, which are as follows.

1. *Insertional vector*   DNA is inserted into a specific restriction enzyme recognition site.

2. *Replacement vector*   Foreign DNA replaces a piece of DNA (stuffer fragment) of the vector.

The replacement vectors are capable of carrying larger DNA inserts. For example, λEMBL4 is a 42 kbp replacement vector that contains 14 of kbp stuffer DNA between the left- and right-hand arms of λ. The joining of λ arms would generate a 28 kbp λ genome, which is too small to be packaged into a λ particle. The suitable packaging size can be attained by the insertion of foreign DNA. So clearly, λEMBL4 is capable of holding foreign DNA fragments up to approximately 23 kbp in size. The size limitations of λ packaging also provide a simple means to ensure the insertion of foreign DNA. Several other basic strategies have also been devised to identify λ phage recombinants, which involves the following.

❖ *Inactivation of the cI gene*   Vectors like λgt11 have unique restriction enzyme recognition sites within the *cI* gene. Upon insertion of foreign DNA, the *cI* gene is inactivated, leading to the formation of clear plaques instead of normal turbid plaques. Screening is dependent on the observer.

❖ *Blue–white screening* Vectors like λZAP contain the *lacZ'* gene expressing the α-fragment of β-galactosidase. Screening is, therefore, similar to that done for pUC plasmids.

M13 VECTORS

M13 is a male-specific (infect only those bacteria harbouring F-pili) lysogenic phage with a circular single-stranded DNA genome (designated as the plus strand) of 6407 base pairs in length. M13 phage particle absorbs through one end to the F-pilus, and the single-stranded phage DNA enters the bacterium and is converted into double-stranded (replicative form, RF) DNA by the synthesis of a complementary DNA strand (the minus strand) using bacterial DNA polymerase. The RF form of the phage genome is rapidly multiplied until about 100 RF molecules are present within the bacterium. Transcription of the viral genes occurs to produce proteins required for the assembly of new viral particles. The production of a virally encoded single-stranded binding protein (the protein product of gene 2) eventually forces asymmetric replication of the RF DNA. This results in only one viral DNA strand being synthesized (the plus strand). These single-stranded DNA molecules are assembled into new viral particles and are released from the cell without the occurrence of cell lysis. Up to 1000 phage particles can be released into the medium per cell per generation. M13 phage infection does not result in bacterial cell death and, consequently, M13 infections appear as turbid plaques.

The M13 origin of replication (called f1 *ori*) has two overlapping, but distinct, DNA sequences that control the initiation and termination of DNA replication with the help of another phage-encoded protein, the product of gene 2. The switch between the double-stranded RF form and the single-stranded plus form of the M13 viral genome has been exploited to use M13 as a vector. M13 vectors have found wide use in site-directed mutagenesis and in DNA sequencing, as discussed earlier.

Unlike λ, M13 does not have a non-essential region that can be deleted prior to the insertion of foreign DNA. The foreign DNA fragments are inserted into an intergenic region between the origin of replication and gene 2. The RF form of M13 vectors can be isolated by standard plasmid DNA preparation procedures, and foreign DNA can be inserted into them as if they were conventional plasmids. The specific use of M13 vectors is as an aid to the formation of single-stranded DNA (Figure 3).

COSMID VECTORS

A cosmid is a plasmid that carries a phage λ *cos* site. Since only a small region in the proximity of the *cos* site is required for recognition by the packaging system, it can be packaged into a phage head. Cosmids infect a host bacterium-like phage but replicate-like plasmids and host cells are not lysed. Cosmids can be used as

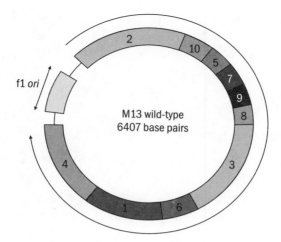

Figure 3   M13 vector

gene-cloning vectors in conjunction with the in vitro packaging system. Cosmid vectors can accommodate 35–50 kb of DNA fragment and are, therefore, widely used for cloning genomic DNAs.

ARTIFICIAL CHROMOSOME VECTORS

Artificial chromosomes are constructed in the laboratory as important tools. An artificial chromosome contains DNA sequences that perform the critical functions of natural chromosomes. They are used to introduce and control new DNA in a cell, to study how chromosomes function, and to map genes in genomes. Bacterial artificial chromosomes (BACs) and yeast artificial chromosomes (YACs) have been largely used in mapping and analysing complex eukaryotic genomes. The complete genome sequencing projects, including the Human Genome Project, largely depended on tools like BACs and YACs, as these vectors are capable of carrying more than 300 kb of foreign DNA fragments.

Bacterial Artificial Chromosomes

BACs are DNA molecules constructed to function as a cloning vector, which can carry 150–350 kb insert and are often modified to carry about 700 kb inserts. A typical BAC vector has the following characteristics.

❖ Antibiotic resistance marker.

❖ Replicon derived from *E. coli* fertility (F) factor and is under stringent control.

❖ An adenosine triphosphate (ATP)-driven helicase (repE) facilitating DNA replication.

❖ Three par loci (parA, parB, parC) to promote accurate partitioning of low-copy number plasmids after host cell division.

For cloning, segments of foreign genomic DNA are ligated to BAC in vitro; the ligation mix is then introduced to recombinant deficient strains of *E. coli* by electroporation. BAC containing foreign DNA becomes established into the host as single-copy plasmids.

BACs are often used in genome sequencing projects for cloning of large overlapping fragments of genomic DNA (Figure 4).

Yeast Artificial Chromosome

These are linear vectors that behave like a yeast chromosome when grown in bacteria or yeast; hence, they are called YACs. YAC vector is capable of carrying up to a million bases of the foreign DNA (Figure 5).

A typical YAC contains the following functional elements from yeast.

❖ An origin of replication, which is an autonomous replication sequence (ARS).

❖ Centromere sequence to ensure segregation into daughter cells (centromeric function).

❖ Telomeric sequences at the two ends, which form hairpin loops at the ends to resist exonuclease action and confer stability.

❖ Growth selectable marker in each arm, for example, URA3, TRP1 genes of yeast. URA3 encodes an enzyme that is required for the biosynthesis of uracil (the nitrogenous base of RNA); TRP1 encodes an enzyme required for the biosynthesis of the amino acid tryptophan.

❖ A selectable marker or integration of foreign DNA insert like SUP4.

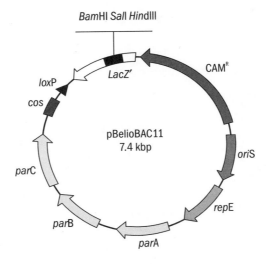

Figure 4 Structure of a BAC vector

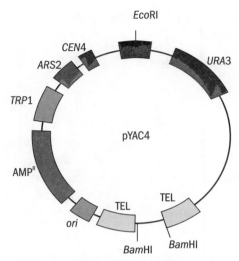

Figure 5 Yeast artificial chromosome vector

❖ The necessary sequences from *E. coli* plasmid for selection and propagation in *E. coli*.

A YAC vector without foreign DNA is maintained as circular DNA. After digestion with *Bam*HI and *Eco*RI, the left arm and right arm become linear with the end telomeric sequences. Foreign DNA cleaved with *Eco*RI is then ligated to the YAC arms and transferred to a *URA3–TRP1* mutant yeast strain as host. The mutant strain requires supplementation with uracil and tryptophan for growth. In recombinant YAC molecules, the arms are joined to bring together the *URA3* and *TRP1* genes, which complement the host, making it capable of growing in nutrient regeneration plates lacking uracil and tryptophan, thus making positive selection through auxotrophic complementation.

Red–white Selection

The multiple cloning site of YAC often contains another marker called SUP4 to facilitate selection of recombinants. SUP4 encodes a tRNA that suppresses the *Ade2-1* UAA mutation. Ade1 and Ade2 code for the enzymes phosphoribosylamino-imidazole-succinocarbozamide synthetase and phosphor-ribosylamino-imidazole carboxylase, respectively. Both of these are involved in adenine synthesis. *Ade2-1* mutant cells, because of the absence of the two enzymes, produce a red pigment, derived from the polymerization of the intermediate phosphoribosylamino-imidazole. But *Ade2-1* mutants expressing SUP4 are like wild type, as *Ade2-1* mutation is suppressed, and are white in colour. When a foreign DNA is inserted in the MSC, the SUP4 is interrupted. In the absence of SUP4 expression, red pigmentation is observed as *Ade2-1* mutation is no longer suppressed. The host strain (*TRP1*- and *URA3*- and *Ade2-1* mutant) containing recombinant YAC produces red colonies, while white colonies are formed when the host contains non-recombinant YAC vector.

Mammalian Artificial Chromosomes

MACs are conceptually similar to YACs, but instead of yeast sequences, they contain mammalian or human sequences. They contain human telomeric repeats of the sequence TTAGGG, and the centromeric sequence is composed of another repeated DNA sequence found at the natural centromeres of human chromosome, the alphoid DNA. When added to suitable cell lines, these MAC DNAs form chromosomes that mimic those in the cell, with accurate segregation and the normal complement of proteins at telomeres and centromeres. Their primary use is delivery of large fragments of DNA to mammalian cells and to whole animals for expression of large genes or sets of genes. They are still under development.

SUMMARY

- Plasmids are widely used as cloning vectors for cloning small DNA fragments. A bacterial plasmid vector must have some properties such as (1) small in size, (2) high copy number, (3) presence of selectable markers, and (4) presence of unique restriction sites for many enzymes so as to be an efficient cloning vector.

- Naturally occurring plasmids are engineered to produce suitable cloning vectors. The plasmids pBR322 and pUC series of plasmids are examples of plasmids that have been engineered for use as cloning vectors.

- The bacteriophages M13 and lambda can be suitably engineered to construct vectors for cloning of DNA fragments. Two basic types of lambda vectors have been developed: insertional vector, wherein DNA is inserted into a specific restriction enzyme recognition site and replacement vector, wherein foreign DNA replaces a piece of DNA (stuffer fragment) of the vector.

- M13 vectors have found wide use in site-directed mutagenesis and in DNA sequencing. The switch between the double-stranded replicative form and the single-stranded plus form of the M13 viral genome has been exploited to use M13 as a vector.

- The cloning of large DNA fragments can be done by using cosmid vectors and artificial chromosome vectors. Artificial chromosome vectors are constructed by engineering bacterial yeast or mammalian chromosomes suitably. Cosmids carry 35–45 kb DNA fragments; 150–350 kb insert can be carried by bacterial artificial chromosomes (BACs), yeast artificial chromosomes (YACs); and mammalian artificial chromosomes (MACs) carry about millions of base pairs of DNA fragment.

- Plasmids are of smaller size compared to chromosomal DNA and occur in covalently closed circular forms. The separation of plasmid DNA from the chromosomal DNA is based on these properties. The isolation procedure involves the growth of bacteria-containing plasmid, lysis of cells by enzyme or alkali or heat treatment, and purification of plasmid DNA.

REVISION QUESTIONS

1. What are cloning vectors? Give an illustrated account of different bacterial plasmids as cloning vectors.
2. Describe the essential features of a plasmid and a lambda phage vector. What are the advantages and applications of plasmids and lambda phage as cloning vectors?
3. What are the advantages of using each of the following vectors as alternatives to plasmids: (1) lambda vector, (2) cosmid vector, (3) M13 phage vector?
4. What are the modifications made in the bacteriophage lambda to generate a typical lambda vector?
5. What is done to prevent the lambda vector from entering into the lysogenic state with the host bacteria?
6. What determines the size range of inserts that can be cloned in a lambda phage vector?
7. What advantages do YAC and BAC have over a plasmid cloning vector? Give examples of YAC and BAC.

15

Isolation and Purification of Plasmid Vectors

OBJECTIVES

After reading this chapter, the student will be able to:
- Explain the classical CsCl-EtBr method
- Discuss alkaline lysis method
- Describe boiling method

INTRODUCTION

One of the prerequisites of the cloning procedure is obtaining the vector DNA in sufficient quantity and in purified form so that the cloning procedure is successful. Plasmid vectors containing bacteria are grown in a medium containing the appropriate antibiotic to ensure that only bacteria-containing plasmids grow. Bacterial cells that do not contain plasmid will be killed in this process of culturing bacteria. Further, there are several methods of extracting and purifying plasmids from bacteria. The basic principle for purification of plasmids takes advantage of the difference between the small, circular plasmid molecules and the large, broken (hence linear) pieces of chromosomal DNA. Chromosomal DNA of bacteria, in most cases, is circular, but being of large size, it is usually broken during isolation, and therefore chromosomal DNA is found in linear pieces. During denaturation, the two strands easily denature and separate. During renaturation, they have difficulty finding complete copies of their complements. Following denaturation, the two circular single-stranded chains of the plasmid DNA remain entwined and do not separate fully. When conditions are such that renaturation can occur, each strand rapidly finds its complement. In an isolation procedure, it is desirable to remove as much chromosomal DNA as possible, at the same time, minimizing the loss and breakage of plasmid DNAs. The isolation procedure involves the growth of bacteria-containing plasmid, lysis of cells, and purification of plasmid DNA.

The starting bacterial culture should be a single transformed colony and should be grown in liquid culture under selective conditions, that is, in the presence of suitable antibiotic. After growth (preferably late logarithmic phase), the bacteria are collected by centrifugation. There are various procedures for lysis, for example, treatment with ionic or non-ionic detergents, lysozyme–ethylenediaminetetraacetic acid (EDTA), alkali, or heat treatment. The choice of reagent depends on the size of the plasmid, the host bacteria, and the method to be used for purification.

THE CLASSICAL CsCl–EtBr METHOD

The classical method is due to Vinograd and involves buoyant density centrifugation in gradients of caesium chloride (CsCl) containing ethidium bromide (EtBr). The dye EtBr binds strongly to DNA, causing the helix to unwind when EtBr intercalates between DNA base pairs. Plasmids, being CCC molecules, unwind to a limited extent and, therefore, bind less EtBr compared to fragmented linear chromosomal DNA. A linear DNA molecule binds EtBr up to saturation. Binding of EtBr causes a decrease in buoyant density for both linear and CCC molecules. Since linear DNA molecules bind more EtBr, they have a lower buoyant density compared to closed circular DNA in CsCl gradient saturated with EtBr. Thus closed circular DNA forms a band at lower position and can be easily separated from the chromosomal DNA (Figure 1). A small needle and a disposable syringe are used to remove the plasmid DNA band from the centrifuge tube. EtBr is then extracted from the plasmid DNA by organic solvent, which is then removed by dialysis. Several modifications of the above procedure have been described. This method yields pure plasmid DNA but is time consuming and expensive. So it is used sparingly.

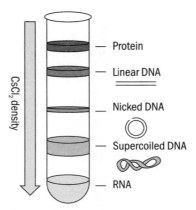

Figure 1 Isopycnic centrifugation of DNA molecules through caesium chloride–ethidium bromide gradient

ALKALINE LYSIS METHOD

Another method, namely, the alkaline lysis method of Birnboim and Doly, is commonly used because of its simplicity, low cost, and reproducibility (Figure 2). This method is based on the fact that at extremely alkaline pH like 12.0–12.5, linear chromosomal DNA is denatured but the covalently closed plasmid DNA is unharmed. In this method, bacterial cells are lysed by a solution containing sodium dodecyl sulphate (SDS) and sodium hydroxide (NaOH). NaOH denatures chromosomal DNA, while SDS denatures bacterial proteins. Upon lowering the pH with sodium acetate, the chromosomal DNA forms an insoluble aggregate. The high concentration of sodium acetate also precipitates protein–SDS aggregate and high molecular weight RNA. The CCC plasmid DNA remains in the aqueous phase and can be recovered and later concentrated by precipitation with ethanol, followed by centrifugation. This method yields plasmid preparations to be used for restriction enzymes digestion, which are suitable for transformation experiments.

Boiling Method

In the boiling method, the bacterial cells are partially lysed with a non-ionic detergent Triton X-100, followed by boiling in the presence of lysozyme. The chromosomal DNA remains attached to the bacterial membrane, and during a brief spin, it is pelleted at the bottom

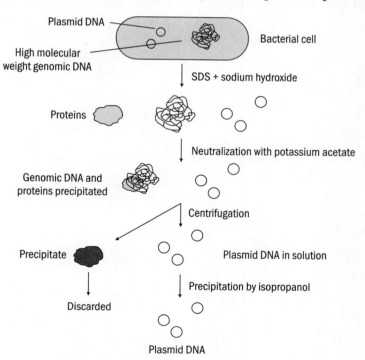

Figure 2   Plasmid DNA preparation by alkali lysis

of the centrifuge tube. The clear supernatant contains the plasmid DNA along with bacterial RNA. Plasmid DNA is precipitated from the supernatant with isopropanol. The produce is less time consuming, but the quality of DNA produced is inferior to that obtained from the alkaline lysis (miniprep). Alternatively, the supernatant containing plasmid DNA may be purified by CsCl–EtBr density gradient centrifugation. However, this procedure leads to low recovery and, in some cases, nicked DNA molecules.

SUMMARY

- One of the prerequisites of the cloning procedure is obtaining the vector DNA in sufficient quantity and also in purified form so that the cloning procedure is successful.
- The basic principle for purification of plasmids takes the advantage of the difference between the small, circular plasmid molecules and the large, broken (hence linear) pieces of chromosomal DNA.
- The classical method of isolation and purification of plasmids is due to Vinograd and involves buoyant density centrifugation in gradients of caesium chloride (CsCl) containing ethidium bromide (EtBr).
- Another method, namely, the alkaline lysis method of Birnboim and Doly, is commonly used because of its simplicity, low cost, and reproducibility. This method is based on the fact that at extremely alkaline pH like 12.0–12.5, linear chromosomal DNA is denatured, but the covalently closed plasmid DNA is unharmed.
- In the boiling method, the bacterial cells are partially lysed with a non-ionic detergent Triton X-100, followed by boiling in the presence of lysozyme.

REVISION QUESTIONS

1. Why is the selective medium containing antibiotics used for culturing bacteria harbouring the plasmid during plasmid isolation?
2. Describe the principle of CsCl–EtBr method of separation of plasmid DNA from the chromosomal DNA.
3. What are the functions of sodium hydroxide and SDS in the cell lysis solution in the alkaline lysis method?
4. What is the function of the sodium acetate solution in alkaline lysis method?
5. What structural property of plasmid DNA allows it to be separated from chromosomal DNA during alkaline cell lysis?

16

Cloning in Cosmid Vectors

OBJECTIVES

After reading this chapter, the student will be able to:
- Describe cosmids
- Learn about the cloning procedure of cosmids
- Discuss the drawbacks, modifications, and rearrangement of cosmids

INTRODUCTION

Cosmids were developed to allow cloning of large pieces of DNA. Cosmids have a capacity of carrying 37–52 kbp of DNA, while normal plasmids are able to carry only 1–20 kbp. Cosmids are conventional plasmids that contain one or two copies of a small region of bacteriophage lambda DNA—the cohesive end site (cos). The cos site contains all of the cis-acting elements required for packaging of viral DNA into bacteriophage lambda particles. Cosmids were developed in light of the observation that concatemers of unit-length lambda DNA molecules can be efficiently packaged if the cos sites are separated by 37–52 kbp intervening sequence. In fact, only a small region in the proximity of the cos site is required for recognition by the packaging system. Hence, in conjunction with the in vitro packaging system, cosmids are used as gene cloning vectors. New phage particles are not produced because cosmids do not carry any of the standard lambda genes, but in vitro packaged phage particles can infect a suitable host.

COSMIDS

Cosmids contain a plasmid origin of replication, a selectable marker, a site into which DNA can be inserted (or in some vectors, multiple cloning site), and a cos site from phage lambda. Figure 1 shows the overall architecture of a pJB8 cosmid vector. Cosmid replicates in different cell types with suitable origin of replication, for example, with

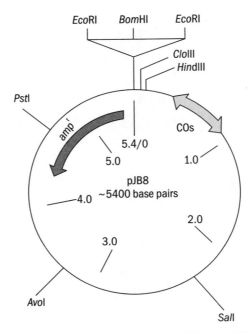

Figure 1 Overall architecture of a pJB8 cosmid vector

SV40 origin of replication, it replicates in mammalian cells; with ColE1 origin, it is suitable for double-stranded DNA replication; and with f1 origin, it is suitable for single-stranded DNA replication. The selectable marker, like plasmids, is usually an antibiotic resistance marker so that the transfected cells are selected by plating in a medium containing the antibiotic. In addition, it may contain *lacZ* marker. Cosmids also possess a unique restriction enzyme recognition site into which DNA fragments can be ligated or in some vectors, multiple cloning sites are present. *Cos* sequences are about 200 base pairs, and these are essential for packaging. There is a *cosN* site where DNA is nicked at each strand, 12 base pairs apart, by terminase (the packaging enzyme). Nicking results in the generation of linearization of the circular cosmid, with two cohesive or sticky ends of 12 base pairs at each termini. The *cos* binding site (*cosB* site) is adjacent to *cosN* and holds the terminase, while it nicks and separates the strands. The *cosQ* site of next cosmid is held by the terminase after the previous cosmid has been packaged, to prevent degradation by cellular DNases. Since the size of the phage head is fixed, terminase can only package cosmids that are between 75% and 105% of the length of the normal phage. This sets the practical upper limit of the insert size.

After the packaging reaction has occurred, the newly formed λ particles are used to infect *Escherichia coli* cells. When λ DNA is injected into the bacterium, it circularizes through complementation of *cos* ends. However, λ infection will not proceed due to the lack of other λ sequences. The circularized DNA will be maintained in the *E. coli* cell as a plasmid. The selection of transformants is made on the basis of antibiotic resistance, and bacterial colonies will be formed, which contain the recombinant cosmid. Since λ phage particles can accept between 37 kbp and 51 kbp of DNA, and most cosmids are about 5 kbp in size, between 32 kbp and 47 kbp of DNA can be cloned into these vectors. This represents considerably more than that could be cloned into a λ vector itself.

CLONING PROCEDURE

The scheme of cloning foreign DNA into a cosmid is illustrated in Figure 2. The vector is cut with a restriction enzyme and then mixed with pieces of foreign DNA to be cloned. Ligation reaction is set up when DNA ligase joins the cut vector and inserts fragments into concatemeric molecules. The ligation mixture is then mixed with a packaging extract. The packaging extracts are essentially lysates of phage-infected *E. coli*. These lysates contain empty phage heads, unattached phage tails, and phage-encoded proteins required for packaging DNA into phage heads.

Whenever two *cos* sites are present on a concatemer and separated by 40 000–50 000 nucleotides, they will be cut and packaged into phage heads. The cosmid-containing phages adsorb to the host, and the cosmid DNA is injected into the cell, which circularizes due to its sticky ends. The annealed ends are then covalently joined by

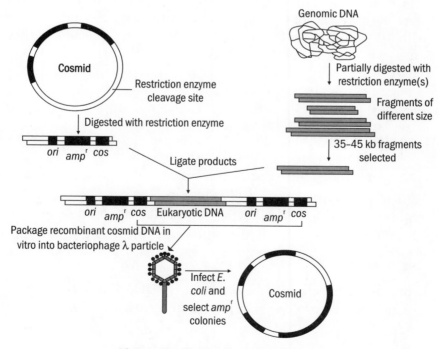

Figure 2 Cloning in cosmid vector

the host's ligase, and the resulting large circular molecule replicates as a plasmid. The selection of transformants is made on the basis of antibiotic resistance, and bacterial colonies (rather than plaques) will be formed, which contain the recombinant cosmid. Since λ phage particles can accept between 37 kbp and 52 kbp of DNA, and most cosmids are about 5 kbp in size, between 32 kbp and 47 kbp of DNA can be cloned into these vectors. This represents considerably more than that could be cloned into a λ vector itself.

Although the use of cosmids for cloning large DNA fragments was a promising approach, a number of problems were encountered, and these were subsequently taken care of by using other strategies or design of refined vectors circumventing a particular problem, as discussed subsequently.

DRAWBACKS AND MODIFICATIONS OF COSMID VECTORS

Rearrangement of Cosmids

Cosmid vectors replicate within cells using their pBR322 origins. Intracellular cosmids sometimes rearrange DNA inserted into them, perhaps because the time taken by the pBR322-dependent replication to replicate the 50000 base-pair cosmid is almost as long as the generation time of *E. coli*, so that cosmids that have deleted sections of DNA have a growth advantage on antibiotic-containing medium. Cosmid vectors (called *lorist vectors*) have recently been developed, which are said to circumvent this problem by using the lambda *ori* and *O* and *P* proteins to replicate. It takes only a few minutes for these vectors to replicate inside cells, and insertions in them are said to be more stable.

SUMMARY

- Cosmids are conventional plasmids that contain one or two copies of a small region of bacteriophage lambda DNA—the cohesive end site *(cos)*.
- In conjunction with the in vitro packaging system, cosmids are used as gene cloning vectors.
- Cosmid replicates in different cell types with suitable origin of replication.
- Cosmids also possess a unique restriction enzyme recognition site into which DNA fragments can be ligated or in some vectors, the multiple cloning sites are present.
- Although the use of cosmids for cloning large DNA fragments was a promising approach, a number of problems were encountered, and these were subsequently

taken care of by using other strategies or design of refined vectors circumventing a particular problem.

REVISION QUESTIONS

1. What are cosmids? Describe the structure of one cosmid vector mentioning the utility of each component.
2. Explain how a foreign DNA can be cloned into a cosmid.
3. How do cosmids rearrange the DNA inserted into them?

17

Construction of Genomic DNA Libraries in Cosmid Vectors

OBJECTIVES

After reading this chapter, the student will be able to:
- Understand cloning into single *cos* site vectors (pJB8)
- Describe cloning into double *cos* site vectors (SuperCos-I)

INTRODUCTION

For the construction of genomic DNA library in cosmid vectors, segments of DNA are ligated in vitro to the vector DNA forming concatemers that are packaged in vitro into bacteriophage lambda particles. These phages when infected in *E. coli* deliver recombinant DNA molecules efficiently into the bacteria where the DNA circularizes and is propagated as a large plasmid. The cosmid vectors are usually of size 5–7 kb; recombinant cosmids can contain not less than 28 kb, and the maximum limit could be 45 kb of DNA insert.

For a successful representation of the entire genome in a cosmid library, the aim should be to cover five to seven equivalents of the target genome. After creation, the library is generally amplified and stored frozen as a single pool of transformed colonies. Cosmid vectors contain one or two *cos* sites. Working with the older single *cos* vectors (for example, pJB8) requires many steps; the newer two *cos* vectors (SuperCos-1) are more convenient to use. Cloning strategies and construction of libraries in these two types of vectors are discussed subsequently.

CLONING INTO SINGLE *COS* SITE VECTORS (pJB8)

The strategy for cloning into single *cos* site vector differs from that adopted for double *cos* site vectors. So these are discussed separately. The cloning into single *cos* site vectors involves the following steps (Figure 1).

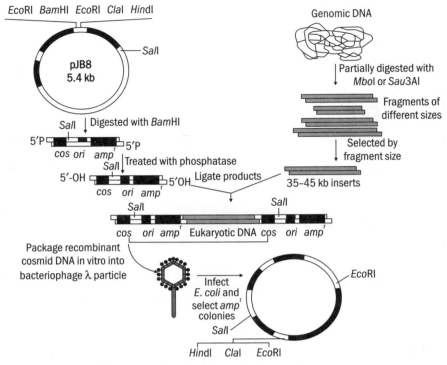

Figure 1 Cloning into single *cos* site vector

❖ Vector digested with *Bam*HI.

❖ Dephosphorylated with alkaline phosphatase, yielding a vector with protruding 5′-termini.

❖ Genomic DNA fragmented by partial digestion with *Mbo*I or *Sau*3AI to generate *Bam*HI compatible ends.

❖ Genomic DNA size fractionated (sucrose density gradient, preparative gel electrophoresis) to retain only 35–40 kb fragments.

❖ Vector and insert ligated and concatemers formed.

❖ Concatemers packaged in vitro into bacteriophage lambda particles.

❖ Infected to *E. coli* and selected by ampicillin resistance.

For cloning into cosmid vectors, genomic DNA fragments of appropriate size are generated by partial digestion of the chromosomal DNA with a restriction enzyme that recognizes a 4 base-pair sequence and generates a cohesive terminus. The enzymes most widely used for this purpose are *Mbo*I and *Sau*3AI. These restriction endonucleases generate fragments with the same cohesive termini as *Bam*HI, and the fragments, therefore, can be cloned into a *Bam*HI site.

Partially digested DNA consists of a population of DNA fragments ranging in size from hundreds of base pairs to over 100 000 base pairs in length. To separate the mixture of DNA fragments into different size classes, two procedures are usually adopted: sucrose density gradient or preparative gel electrophoresis. Size fractionation by sucrose density gradient is accomplished by heating the DNA solution to dissociate aggregated DNA fragments and then loading it onto a high-salt sucrose gradient. After centrifugation and gradient fractionation, the appropriate fractions are identified by agarose gel electrophoresis. The digested genomic DNA can be size fractionated on a slab agarose gel. After fractionation, the appropriate region of the gel is defined by size, and the DNA is eluted. DNA fragments of 35–45 kb are chosen for further ligation reaction.

The DNA of cosmid pJB8 is digested with *Bam*HI and dephosphorylated with alkaline phosphatase to yield a vector with protruding 5'-termini.

A ligation reaction is set up after optimizing the conditions of the reaction. For optimization, a number of small-scale ligation reactions are set up using fixed amount of vector and varying amount of insert. Test ligations are packaged and plated on host bacteria. The number of clones in the different ligations is compared, and the optimum ratio of vector to insert is indicated by the ligation with the most recombinant clones.

The resulting concatemers are ready for in vitro packaging of bacteriophage lambda particles. In vitro packaging uses lysates of phage-infected *E. coli*, called packaging extracts. These lysates contain empty phage heads, unattached phage tails, and the phage-encoded proteins required for DNA packaging. If adenosine triphosphate (ATP) is present, concatemeric phage DNA mixed with the extract is cut at one *cos* site by the terminase (probably a complex of the *A* and *Nul* proteins) and loaded into the phage head by an unknown mechanism. DNA continues to be loaded into the phage head until the terminase encounters and cuts the next *cos* site on the molecule. The phage tails then attach themselves to the filled heads.

When introduced into *E. coli*, the cosmid DNA recircularizes and replicates in the form of a large plasmid. The plasmid contains a β-lactamase gene that confers resistance to ampicillin on the host bacterium.

CLONING INTO DOUBLE *COS* SITE VECTORS (SuperCos-1)

Cloning into a vector containing double *cos* sites involves the following steps (Figure 2).

❖ Vector digested with *Xba*I to linearize.

❖ Treated with alkaline phosphatase for dephosphorylation.

❖ Further digested with *Bam*HI to separate *cos* segments that are carried onto separate fragments.

❖ Genomic DNA partially digested with *Mbo*I.

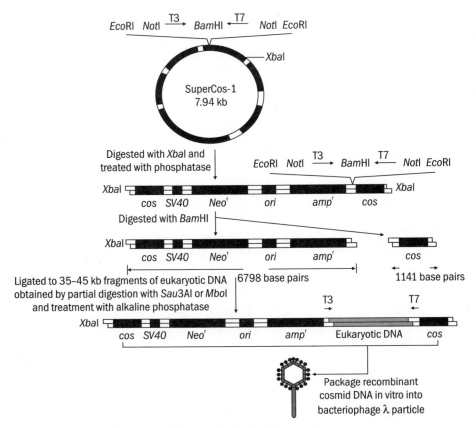

Figure 2 Cloning into double *cos* site vector

❖ Genomic DNA treated with alkaline phosphatase for dephosphorylation; alternative strategy may be adopted at this stage for dephosphorylation.

❖ Vector and inserts ligated, genomic DNA fragments carry termini that are compatible with *Bam*HI cohesive termini of the vector.

❖ The concatenated DNA is then packaged into bacteriophage lambda particles that are used to infect a *recA*⁻ strain of *E. coli*.

In double *cos* site vectors, the two *cos* sites are separated by a recognition site for a restriction enzyme (in this case *Xba*I) that cleaves the vector only once. The vector arms are prepared by first digesting the DNA with *Xba*I and then removing the 5′-phosphate residues from the linearized cosmid by treatment with alkaline phosphatase. In a second digestion, the linearized double *cos* vector is digested with *Bam*HI to produce two arms, each of which carries a *cos* site. Although, at best only 50% of the concatemers can have the correct arrangement of *cos* sites, such molecules are packaged into bacteriophage lambda heads with very high efficiency, provided they are between 35 kb and 52 kb in length. This generates about 10^5 to 10^7 transformed colonies per microgram of DNA.

For standardization of partial digestion, genomic DNA is treated with limiting amounts of the particular restriction enzyme for variable lengths of time. At different time intervals, the samples are withdrawn and analysed by agarose gel electrophoresis to determine the average length of the digested DNA. The combination of time points and concentration of enzymes used in digestion, which gives most enriched products for the desired size fractions, are then used as guide for the preparative digestion of the same DNA. In this way, size-selected DNA fragments can be prepared with a minimum of trial and error.

The genomic DNA is partially digested with *Mbo*I so that the fragments carry termini that are compatible only with the *Bam*HI cohesive termini of the vector. The partially digested genomic DNA is dephosphorylated by alkaline phosphatase, as described above.

The two arms are then ligated to partially digested and dephosphorylated genomic DNA generating, inter alia, molecules in which two *cos* sites are oriented in the same manner and separated by genomic DNA inserts.

The concatenated DNA is then packaged into the bacteriophage lambda particles, which are used to infect a *recA*⁻ strain of *E. coli*.

SUMMARY

- The strategy for cloning into single *cos* site vectors differs from that adopted for double *cos* site vectors.

- For cloning into cosmid vectors, genomic DNA fragments of appropriate size are generated by partial digestion of the chromosomal DNA with a restriction enzyme that recognizes a 4 base-pair sequence and generates a cohesive terminus. The enzymes most widely used for this purpose are *Mbo*I and *Sau*3AI.

- In double *cos* site vectors, the two *cos* sites are separated by a recognition site for a restriction enzyme (in this case *Xba*I) that cleaves the vector only once.

REVISION QUESTIONS

1. Why partial digestion is required for the construction of genomic libraries?
2. What are the advantages of double *cos* site vectors over the single *cos* site vectors?
3. How is a desired gene isolated from the cosmid library?

18

Enzymes Used in Molecular Cloning

OBJECTIVES

After reading this chapter, the student will be able to:

- Describe *E. coli* DNA polymerase I (holoenzyme)
- Discuss Klenow fragment of DNA polymerase I
- Explain T4 DNA polymerase
- Describe T7 DNA polymerase
- Understand thermostable DNA-dependent DNA polymerases
- Explain terminal deoxynucleotidyl transferase
- Describe reverse transcriptase
- Discuss phage RNA polymerases: T7, T3, SP6
- Explain poly(A) polymerase
- Analyse alkaline phosphatases
- Describe T4 polynucleotide kinase
- Discuss exonucleases, endonucleases, and BAL 31 nuclease enzyme
- Describe S1 nuclease

INTRODUCTION

The major tools for genetic engineering are the enzymes that catalyse specific reactions on DNA molecules. The utility of many of these enzymes has been discussed earlier. This chapter reviews the principal enzymes critical for carrying out most of the important reactions involved in recombinant DNA technology. The chapter also explains the digestion procedure of DNA fragments using restriction enzymes.

E. coli DNA POLYMERASE I (HOLOENZYME)

DNA polymerase I (Pol I), discovered by Kornberg in 1956, is encoded by *Escherichia coli polA* gene and is composed of a single polypeptide chain of molecular mass of about 103 kDa. DNA Pol I possesses three enzymatic activities, which are as follows.

❖ 5′→3′ DNA polymerase activity, requiring a 3′-primer site and a template strand.

❖ 3′→5′ exonuclease activity that mediates proofreading.

❖ 5′→3′ exonuclease activity mediating nick translation during DNA repair.

polA gene has been cloned and sequenced. It codes for 928 amino acids protein. The gene has been expressed from prokaryotic high expression vector, and it is possible to isolate large quantities of protein for commercial use. *polA* contains the following domains.

❖ The DNA polymerase and 3′→5′ exonuclease activities are carried on the largest domain present on the C-terminal (amino acids 543–928). The uses and applications of Klenow fragment are discussed later.

❖ The 5′→3′ exonuclease activity of the holoenzyme is carried on the smaller amino terminal fragment (residues 1–325). This also performs a proofreading function by resecting DNA that contains either mismatched bases or frame-shift errors. The 3′→5′ exonuclease of *E. coli* DNA Pol I is less active than that of T4 or T7 DNA polymerase.

Applications

❖ *Uniform labelling of DNA by nick translation* The mechanism of nick translation involves the combined activities of the 5′→3′ polymerase and 5′→3′ exonuclease of a DNA-dependent DNA polymerase, usually *E. coli* DNA Pol I.

KLENOW FRAGMENT OF DNA POLYMERASE I

The 5′→3′ exonuclease activity of DNA polymerase I made it unsuitable for certain applications. Treatment of DNA Pol I by protease subtilisin cleaves the molecule into a small N-terminal part (1–325 amino acids) retaining the 5′→3′ exonuclease activity and a 76 kDa large C-terminal fragment called Klenow fragment. It possesses the DNA polymerase and 3′→5′ exonuclease activity of the parent Pol I enzyme, but lacks the 5′→3′ exonuclease activity and has wide applications in molecular biology (Figure 1).

The Klenow fragment has been cloned and expressed from a high expression vector to yield a large amount of this fragment for several applications. In some situations, the 3′→5′ exonuclease activity of Klenow fragment is not required or undesirable. A mutated form of the Klenow fragment is available. This enzyme retains the polymerase activity but lacks the 3′→5′ exonuclease activity.

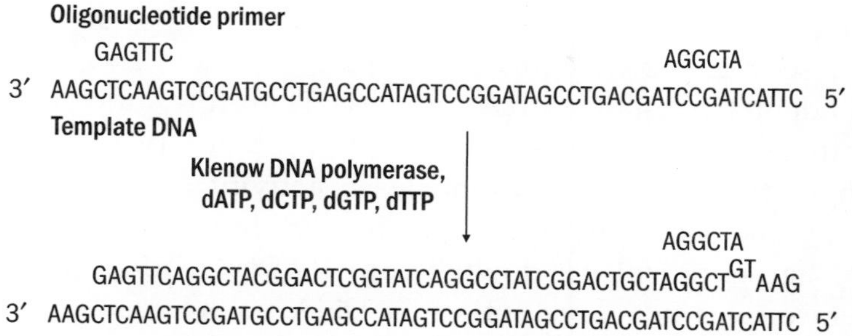

Oligonucleotide primer

GAGTTC   AGGCTA

3′ AAGCTCAAGTCCGATGCCTGAGCCATAGTCCGGATAGCCTGACGATCCGATCATTC 5′

Template DNA

Klenow DNA polymerase,
dATP, dCTP, dGTP, dTTP

AGGCTA

GAGTTCAGGCTACGGACTCGGTATCAGGCCTATCGGACTGCTAGGCT^{GT}AAG

3′ AAGCTCAAGTCCGATGCCTGAGCCATAGTCCGGATAGCCTGACGATCCGATCATTC 5′

Figure 1   Klenow DNA polymerase

Applications

❖ *Labelling of DNA by random oligonucleotide-primed synthesis*   Random oligonucleotide-primed synthesis is a procedure for producing uniformly radioactive DNA of high specific activity. The DNA is cleaved with a restriction endonuclease, denatured by heating at 95°C, annealed to random-sequence oligodeoxynucleotides (typically six bases in length), and then incubated with the Klenow (3′→5′ exo-) fragment in the presence of deoxynucleotide triphosphates (dNTPs). In this way, the hexanucleotides prime the DNA of interest at various positions along the template and are extended to generate double-stranded DNA, which is uniformly labelled on both strands.

❖ *Repairing 3′ or 5′ overhanging ends to generate blunt ends*   For many cloning experiments, blunt end products are required. The enzymes necessary to convert the ends generated by restriction T4 or native T7 DNA polymerases can efficiently chew up 3′ extensions. Klenow is preferred in many cases due to its lower exonuclease/polymerase activity ratio, which allows room temperature incubations. It can repair or chew up the ends generated by restriction endonucleases and other exonucleases like Bal 31, λ exonuclease, and exonuclease III or endonucleases, such as S1 or mung bean nuclease.

Other applications of Klenow fragment are as follows.

❖ DNA sequencing by the Sanger dideoxy method.

❖ Synthesizing the second strand for cloning of cDNAs.

❖ Extending oligonucleotide primers on single-stranded templates. Some specific applications include dideoxy DNA sequencing, synthesis of hybridization probes, and in vitro mutagenesis.

❖ Converting single-stranded oligonucleotides to double-stranded DNA by mutually primed synthesis.

T4 DNA POLYMERASE

T4 DNA polymerase, encoded by gene 43 of bacteriophage T4, is a monomeric protein of 104 kDa. It is produced either from *E. coli* cells that have been infected with the phage or from the cloned and overexpressed gene in *E. coli*. T4 DNA polymerase activities are very similar to those of Klenow fragment. It has a 3′→5′ exonuclease activity in addition to its DNA-dependent DNA polymerase activity. There are, however, two differences of practical significance. First, the 3′→5′ exonuclease activity of T4 DNA polymerase is 200-fold stronger than the 3′→5′ exonuclease activity of Klenow fragment, which makes T4 DNA polymerase preferable in producing blunt end DNA from 3′ overhang. Second, Klenow fragment displaces downstream oligonucleotides as it polymerizes, but T4 DNA polymerase does not. This makes T4 DNA polymerase an efficient choice for oligonucleotide mutagenesis reactions.

Applications

❖ *Repairing 3′ or 5′ overhanging ends to generate blunt ends*   This application is very similar to that described for Klenow fragment. However, compared to Klenow, T4 DNA polymerase requires a higher concentration of dNTPs and a lower reaction temperature. Repair of 5′ extensions is carried out by polymerase activity, whereas repair of 3′ extensions is carried out by 3′→5′ exonuclease activity.

Other applications of T4 DNA polymerase are as follows.

❖ Radioactive labelling of the 3′-termini of DNA fragments.

❖ Selective and extensive labelling of the 3′-termini of a linear duplex DNA molecule.

❖ Gap filling by utilizing the 5′→3′ polymerase activity to extend the 3′-OH end upstream of the gap to the downstream 5′-phosphate end. The remaining nick can be sealed by a variety of different DNA ligases.

❖ Converting the ends of any duplex DNA fragment to blunt ends suitable for blunt-end ligation and subsequent cloning.

T7 DNA POLYMERASE

T7 DNA polymerase is synthesized in *E. coli* cells following infection with the bacteriophage T7. T7 polymerase is a complex of two tightly bound proteins in a one-to-one stoichiometric amount. These are the T7-encoded gene 5 protein (80 kDa) and the *E. coli* encoded thioredoxin (12 kDa). Both these proteins have been overproduced in *E. coli* from genes cloned into plasmid vectors. This complex or the T7 polymerase is considered the most processive of all known DNA polymerases. In other words, the average length of DNA synthesized before the enzyme dissociates from the template is considerably higher compared to other polymerases. Purified T7 gene 5 protein has a 3′→5′ exonuclease activity and a non-processive DNA polymerase activity. Thioredoxin acts as an accessory protein to increase the affinity of T7 gene 5 protein for the primer template, rendering DNA synthesis processive for thousands of nucleotides. Due to its high processivity, this enzyme finds its use in sequencing by Sanger's method.

However, the native T7 DNA polymerase has a very active single- and double-stranded DNA 3′→5′ exonuclease activity, which is often a detriment to the use of this enzyme for DNA sequence analysis. So T7 DNA polymerase is either selectively reduced by a chemical reaction with FeSO or genetically engineered (deletion of 28 amino acids from the exonuclease domain of the enzyme) to get rid of its exonuclease activity. This form of enzyme is marketed under the name Sequenase and Sequenase 2.0.

Applications

❖ *DNA sequencing*   T7 DNA polymerase is highly processive and can be used for extensive synthesis of DNA on long templates. It is largely unaffected by secondary structures that impede *E. coli* DNA pol I, T4 DNA polymerase, or reverse transcriptase.

❖ Native T7 DNA polymerase is the enzyme of choice for synthesizing the complementary strand during site-directed mutagenesis.

❖ Native T7 DNA polymerase can be used analogously to T4 DNA polymerase to convert the protruding ends of any duplex DNA fragment to blunt ends.

❖ Primer extension reaction that requires the copying of long stretches of template.

❖ Rapid end labelling by either filling or replacement reactions.

THERMOSTABLE DNA-DEPENDENT DNA POLYMERASES

Thermostable DNA-dependent DNA polymerases have been isolated and purified from thermophilic and hyperthermophilic organisms, which may be eubacteria or archaea. Although there are considerable structural differences among the various thermostable DNA polymerases, all of them are monomeric and have molecular weight ranging from 80 kDa to 100 kDa.

The most well-known thermostable DNA polymerase was purified from the bacterium *Thermus aquaticus*, which was found in hot spring and belonged to extreme thermophilic group and was called *Taq* DNA polymerase. This study was published in 1976, and 10 years later, polymerase chain reaction (PCR) was developed. Soon thereafter, the *Taq* DNA polymerase became indispensable in molecular biology laboratory. *Taq* DNA polymerase has been cloned in *E. coli* and is now widely available for many commercial sources. The enzyme is a single 94 kDa polypeptide. Like other DNA polymerases, it catalyses the template-directed synthesis of DNA from nucleotide triphosphates. A primer having a free 3′-OH end is required to initiate synthesis in the presence of Mg^{2+} ions. It possesses a 5′→3′ structure-dependent nuclease activity, in addition to its 5′→3′ polymerase activity. *Taq* DNA polymerase has no 3′→5′ exonuclease activity. In general, *Taq* DNA polymerase has optimum activity at 75–80°C, but the activity is reduced at lower temperature; at 60°C, about 50% activity is retained, while at 37°C, only about 10% activity is retained.

The N-terminus of *Taq* polymerase can be genetically deleted, leaving behind a 61 kDa form called the Stoffel fragment. The Stoffel fragment has a slightly enhanced level of thermostability but is devoid of 5′→3′ nuclease activity and is useful when working with DNA templates containing significant secondary structure, which require higher denaturation temperatures.

The discovery of *Taq* polymerase and its immense application in the PCR technology led to the search of other thermostable DNA polymerases. In addition to *Taq* DNA polymerase, several others have been isolated with enhanced processivity or higher levels of fidelity or thermostability and have further advanced this field. These are Vent DNA polymerase, Deep Vent DNA polymerase, and *Pfu* DNA polymerase. Vent DNA polymerase was isolated from the archaea *Thermococcus litoralis*, which have a 3′→5′ proofreading exonuclease activity, resulting in high fidelity DNA synthesis. Deep Vent and *Pfu* DNA polymerases also have still higher levels of fidelity and thermostability.

One of the important criteria considered for use of thermostable DNA polymerases concerns the error rate. As would be expected, polymerases lacking 3′→5′ exonuclease activity generally have higher error rates than polymerases with the above exo-activity. The reports of error rate of *Taq* DNA polymerase from various sources differ as the rates are measured by different assays, but are of the order of 1×10^{-4} to 2×10^{-5} errors per base pair. Pfu polymerase appears to have the lowest error rate at roughly 1.5×10^{-6} error per base pair. Vent polymerase is the intermediate of the two.

TERMINAL DEOXYNUCLEOTIDYL TRANSFERASE

Terminal deoxynucleotidyl transferase (TdT), also called terminal transferase, is a template-independent DNA polymerase that catalyses the addition of deoxynucleotides to the 3′-OH terminus of DNA, accompanied by the release of inorganic phosphate. TdT acts on single-stranded DNA, including 3′ overhangs of double-stranded DNA.

TdT was once purified from calf thymus but is now widely available in recombinant form. The polymerase domain is located at the C-terminus of this 58.3 kDa protein. TdT does not contain a 5′ or 3′ exonuclease domain. Most suppliers provide an N-terminal truncated form of the protein with a molecular weight between 40 kDA and 56 kDa. TdT activity requires divalent cations, and the identity of the cation can influence the fact regarding which nucleotide will be preferentially incorporated (Figure 2).

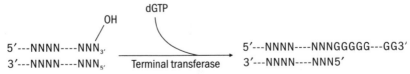

Figure 2 Tailing with terminal transferase

Applications

❖ *Cloning DNA fragments*   TdT is used to synthesize a homopolymer "tail" (for example, tail of Gs) at each end of a linearized vector and a complementary homopolymer tail (Cs) at each end of the DNA to be cloned. When incubated together, the complementary Gs and Cs annealed leads to the annealing of the vector and insert DNAs.

❖ Labelling the 3'-terminus of DNA with [^{32}P]dNTPs.

❖ Synthesizing model polydeoxynucleotide homopolymers.

❖ Detecting apoptosis or DNA damage using the TUNEL, or terminal deoxynucleotidyl transferase-mediated dUTP nick end-labelling assay.

REVERSE TRANSCRIPTASE

It is involved in the conversion of RNA to DNA. It is found in retroviruses where it is responsible for the conversion of their RNA genome into a DNA copy prior to integration into host cells. The enzyme was first discovered by David Baltimore and Howard Temin independently in 1970 for which they received the Nobel Prize. Reverse transcriptase, like all other DNA polymerases, catalyses the addition of new nucleotides to a growing chain in a 5' to 3' direction. Reverse transcriptases generally have two types of enzymatic activity, namely, DNA polymerase and RNaseH activity. In molecular biology laboratory, reverse transcriptase is mainly used as RNA-directed DNA polymerase in first-strand cDNA synthesis. Specifically, oligo-dTs are used as primers for the extension on RNA templates. The DNA synthesized from an RNA template is referred to as complementary DNA (cDNA) and is often used as a template for PCR or converted to double-stranded DNA for cloning. For cDNA synthesis from bacterial mRNAs, the reaction is primed by random primers, usually mixed hexamers. RNaseH activity degrades the RNA from RNA–DNA hybrids, such as those formed during reverse transcription of an RNA template.

The commercially available enzymes used in cDNA library construction are derived either from Moloney murine leukaemia virus (MMLV-RT or MMuLV-RT) or from avian myeloblastosis virus (AMV-RT). The AMV enzyme is purified from isolated virus, while MMLV enzyme is purified following overproduction of the cloned gene in *E. coli* cells. Both enzymes have the same fundamental activities but differ in a number of characteristics, including temperature and pH optima. AMV-RT has a higher optimal reaction temperature (42–45°C) and is preferred for copying RNA with significant secondary structure. MMLV-RT is a single polypeptide of 71 kDa in size, while AMV-RT is composed of two polypeptide chains, 64 kDa and 96 kDa in size. AMV-RT also has a higher processivity, but its fidelity is approximately half that of MMLV-RT. Most importantly, MMLV-RT has a very weak RNaseH activity compared to AMV-RT, which gives it an obvious advantage when being used to synthesize DNA from long RNA molecules.

A mutant reverse transcriptase enzyme with reduced RNaseH activity has been constructed and is commercially available as RNaseH-RT enzymes. In cases when very low amounts of RNA template are used, premature cleavage of templates can be reduced by the use of an RNaseH-RT mutant.

Applications

❖ Synthesizing cDNA as a substrate for downstream PCR or qPCR applications, or for insertion into bacterial cloning vectors, as in the construction of cDNA libraries.

❖ Filling in and labelling the 3'-terminus of DNA fragments with 5'-protruding ends.

❖ For DNA sequencing, in place of Klenow fragment.

PHAGE RNA POLYMERASES: T7, T3, SP6

The most commonly used DNA-dependent RNA polymerases for in vitro transcription reactions include the related bacteriophage T7, T3, and SP6 polymerases.

Earlier studies employed *E. coli* RNA polymerase. It is a large multi-subunit enzyme (~480 kDa) that initiates transcription following recognition of the −10 and −35 sequence elements within a large (~40 base pairs) promoter region. However, it was problematic for many reasons, namely, premature transcription termination in vitro and the inability to generate a uniform population of RNA. Further, in the 1980s, a different class of DNA-dependent RNA polymerases was available; these T7, T3, and SP6 RNA polymerases were of bacteriophage origin and are now preferred to *E. coli* RNA polymerase. These enzymes are encoded by *gene 1* from a family of related bacteriophages. Each of this is a single-subunit enzyme (~95–98 kDa) that recognizes and initiates transcription specifically and exclusively from its own 18 base-pair promoter sequence.

Phage T7, T3, and SP6 RNA polymerases are extremely processive enzymes used for high-yield transcription of DNA sequences inserted downstream of the corresponding T7, T3, or SP6 promoter. Transcripts, thousands of nucleotides in length, are readily obtained without the enzyme dissociating from the template.

Applications

❖ For generating homogeneous single-stranded RNA probes that are uniformly labelled to a high specific activity. Probes generated with phage polymerase are useful for detection of homologous DNA or RNA sequences by standard Southern or Northern hybridization techniques.

❖ For mapping the ends of RNA or DNA using the ribonuclease protection assay.

❖ For generating large quantities of small RNA (for example, tRNA, microRNA, or miRNA, ribozymes) that can be used for physical and biochemical studies.

❖ For generating transcripts that can be used as precursor RNAs for studies on RNA splicing and processing.

POLY(A) POLYMERASE

E. coli poly(A) polymerase catalyses the incorporation of adenosine monophosphate (AMP) residues in a template-independent manner onto the free 3'-OH terminus of RNA, utilizing ATP as a precursor. The enzyme has a strong preference for ATP compared to the other nucleoside triphosphates (NTPs). The length of the poly(A) tail generated can be modulated by adjusting the incubation time, amount of enzyme, and number of 3'-termini available for addition. In the presence of high salt (≥ 250 mM NaCl), processivity of the enzyme is reduced, and the distribution range of the number of AMP residues added to each substrate molecule is decreased.

ALKALINE PHOSPHATASES: BACTERIAL ALKALINE PHOSPHATASE AND CALF INTESTINE PHOSPHATASE

Alkaline phosphatase is a hydrolase enzyme that catalyses the hydrolysis of 5'-phosphate residues from DNA, RNA, and ribonucleoside and deoxyribonucleoside triphosphates. The process of removing phosphate groups is called dephosphorylation.

The removal of the phosphate group allows radiolabelling; the dephosphorylated products possess 5'-OH termini, which can subsequently be radioactively labelled using $[\gamma\text{-}^{32}P]$ ATP and T4 polynucleotide kinase. Also removal of phosphate prevents the DNA from ligation. Usually, the vector molecules are treated with alkaline phosphatase before ligation so that the exposed 5'-phosphate group is removed from the vector DNA. Since DNA ligase requires the presence of a 5'-phosphate group, the vector molecules become unsuitable for ligation (both intramolecular and intermolecular). This prevents self-ligation of vectors. Ligation of vector to insert can occur, as 5'-phosphate groups of the inserts are still exposed. As the name suggests, the enzyme alkaline phosphatase is effective in the alkaline environment and requires Zn^{2+} for activity (Plate 1).

Another enzyme, calf intestine phosphatase (CIP) from veal, is also commonly used in nucleic acid research, which functions like alkaline phosphatase. The primary difference between them is the stability of the two enzymes. CIP is readily inactivated by heating to 70°C for 10 min and/or extraction with phenol. On the other hand, bacterial alkaline phosphatase (BAP) is much more resistant to these treatments. Thus for most purposes, CIP is the enzyme of choice. Furthermore, CIP has a 10- to 20-fold higher specific activity than BAP. Alkaline phosphatase is the enzyme that removes a 5'-phosphate group from many types of molecules, including a DNA fragment. The process of removing the phosphate group is called dephosphorylation. The alkaline phosphatase enzyme is most effective in an alkaline environment.

Applications

❖ Dephosphorylation of 5'-termini of nucleic acids prior to labelling with [γ-^{32}P] ATP and T4 polynucleotide kinase. 5'-^{32}P end-labelled DNA is used for sequencing by the Maxam–Gilbert procedure for RNA sequencing by specific RNase digestions, and in mapping studies using specific DNA or RNA fragments.

❖ Dephosphorylation of 5'-termini of vector DNA in order to prevent self-ligation of vector termini.

T4 POLYNUCLEOTIDE KINASE

T4 polynucleotide kinase, encoded by the phage T4 *pseT* gene, was originally purified from T4-infected *E. coli* cells. Recently, the *pseT* gene has been cloned into *E. coli*, and the enzyme has been overproduced from this strain.

Polynucleotide kinase catalyses the transfer of a phosphate group from ATP to the 5'-ends of a polynucleotide DNA or RNA. The enzyme is used in two types of reactions—forward and exchange.

In the forward reaction, T4 polynucleotide kinase catalyses the transfer of the terminal gamma phosphate of ATP to the 5'-end of DNA and RNA. This reaction is very efficient and, hence, is the general method for labelling 5'-ends or for phosphorylating oligonucleotides.

In the exchange reaction, T4 polynucleotide kinase catalyses the exchange of 5'-terminal phosphates. In this reaction, target DNA or RNA that has a 5'-terminal phosphate is incubated with an excess of ADP; the 5'-terminal phosphate is transferred by polynucleotide kinase to ADP forming an ATP, leaving a dephosphorylated substrate, which is subsequently rephosphorylated in a forward reaction by the transfer from the γ phosphate of [γ-^{32}P] ATP. The exchange reaction is less efficient than the forward reaction; thus it is rarely used (Figure 3).

Finally, polynucleotide kinase is a 3'-phosphatase. Some commercial preparations of polynucleotide kinase are developed from the phage T4 strain am N81 *pseT*1, which has a mutated *pseT* gene. This derivative lacks the 3'-phosphatase activity.

Applications

❖ DNA sequencing by the chemical degradation (Maxam–Gilbert) technique.

❖ Defining specific protein–DNA interactions by DNase I footprinting or protection from DNA-damaging chemicals, such as dimethyl sulphate or methidiumpropyl-ethylenediaminetetraacetic acid (EDTA) (MPE).

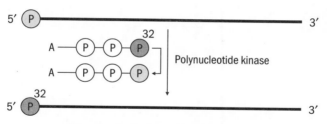

Figure 3 Polynucleotide kinase

❖ Mapping of restriction sites by partial digestion of 5′-end-labelled DNA fragments.

❖ Mapping the termini of RNA transcripts.

❖ Mapping the position of introns in DNA.

❖ Synthesis of substrates for the assay of DNA and RNA ligases.

❖ Labelling oligonucleotides for purification by gel electrophoresis.

❖ Ligation of oligonucleotides into DNA vectors; when oligonucleotides are chemically synthesized, they have 5′-OH termini; for cloning purposes, it is often desirable to phosphorylate them.

❖ Phosphorylation of linkers and adapters or fragments of DNA as a prelude to ligation, which requires a 5′-phosphate.

❖ Radiolabelling of oligonucleotides for purification by gel electrophoresis; application in DNA sequencing by the Maxam–Gilbert method; restriction mapping by partial digestion; mapping the termini of RNA transcripts; mapping the positions of introns in DNA.

EXONUCLEASES

Exonucleases are enzymes capable of cleaving nucleotides one at a time from the end of a polynucleotide chain. The phosphodiester bonds are broken either from the 3′-terminus (in 3′ to 5′ direction) or from 5′-end (5′ to 3′ direction) by a hydrolysing reaction. Exonucleases are often involved in proofreading activity; they remove the sugar–phosphate backbone from the 3′-end in cases wherein the bases are not properly paired (other than A–T or G–C), damaged or missing. Exonucleases were discovered long back in 1964. In 1971, exonuclease I was found by Lehman IR in *E. coli*. Since then, numerous exonucleases have been discovered, each with a specific function and/ or requirement (Figure 4).

ENDONUCLEASES

Endonucleases, on the other hand, cleave phosphodiester bonds in the middle of a polynucleotide chain. This cleavage can be general or sequence specific (restriction endonucleases). Some endonucleases are specific for cutting single-stranded DNA.

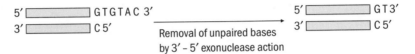

Figure 4 Exonuclease activity

BAL 31 Nuclease Enzyme

Bal 31 nuclease from *Alteromonas espejiana* is a single-strand-specific endodeoxyribonuclease. On duplex circular DNA, it degrades at nicks or at transient single-stranded regions created by supercoiling. On duplex linear DNA, it degrades from both the 5′- and 3′-termini at both ends, resulting in a controlled shortening of the DNA. Bal 31 also acts as a ribonuclease, catalysing the hydrolysis of ribosomal and tRNA. Bal 31 nuclease requires both Ca^{2+} and Mg^{2+}. Digestions can be terminated with ethylene glycol tetraacetic acid (EGTA), a specific chelator of Ca^{2+}, without affecting the Mg^{2+} concentration. Bal 31 nuclease is active in sodium dodecyl sulphate (SDS) and urea.

Applications

❖ Cloning for creating deletions of different sizes in a controlled manner. The cloned DNA is resected at specific sites with a restriction enzyme, and the resulting linear DNA is digested with Bal 31 nuclease for various intervals of time.

❖ Mapping restriction sites in a DNA fragment.

❖ Investigating secondary structure of supercoiled DNA and alterations in the helix structure of duplex DNA, caused by treatment with mutagenic agents.

❖ Creating successive deletions needed for linker-scanning mutagenesis.

S1 NUCLEASE

S1 nuclease from *Aspergillus oryzae* is a highly specific single-stranded endonuclease. The rate of digestion of single-stranded DNA is 75000 times faster than that of double-stranded DNA. The pH optimum is 4.0 to 4.3, and the rate drops 50% at pH 4.9. Reactions are normally run at pH 4.6 to avoid depurination of the DNA. The enzyme requires low levels of Zn^{2+}, and it is largely unaffected by NaCl concentrations in the range of 10–300 mM. It is stable to urea, SDS, and formamide.

Applications

Most applications of S1 nuclease make use of its ability to trim protruding single-stranded ends of DNA and RNA without significant nibbling of blunt duplex ends.

Some of these applications are as follows.

❖ Mapping the 5'- and 3'-ends of RNA transcripts by the analysis of S1-resistant RNA–DNA hybrids.

❖ Mapping the location of introns by digesting a hybrid of mature mRNA with ^{32}P-labelled genomic DNA. S1 will cleave at the single-stranded loops created by introns within these hybrid molecules.

❖ Digesting the hairpin structures formed during the synthesis of cDNA by reverse transcriptase.

❖ Removing single-stranded termini of DNA fragments to produce blunt ends for ligation.

❖ Creating small deletions at restriction sites. It will nibble at the ends of a linear duplex fragment at a very slow rate.

❖ Rendering the ends flush after successive deletions by exo III for linker-scanning mutagenesis.

Besides the above description, there are several other enzymes used in different molecular biology techniques. Some of the important enzymes are described in Table 1. All the enzymes discussed in Table 1 are summarized in Table 2.

RESTRICTION ENZYMES: DIGESTION OF DNA WITH RESTRICTION ENZYMES

In the simplest case, for example, digestion of a single DNA sample with a single restriction enzyme, simple incubation of the DNA sample with enzyme is required. The amount of DNA, enzyme, buffer, temperature, and duration of incubation are decided from manufacturer's instructions according to the experimental requirement. When digestion of DNA with multiple restriction enzymes is required, the buffer in which these enzymes are active is the most important criteria. If, for example, we need to digest a DNA fragment with two restriction enzymes, which can be used with similar buffer, then the case is simple and any one buffer can be used. Some enzymes are active at a wide range of salt concentrations. However, if one enzyme (A) is active at low salt concentration, while the other (B) is active at high salt concentration, digestion may be done sequentially, first with A and then with B. Similar procedure can be adopted in case the temperatures of incubation of two enzymes are different.

Partial Digestion

Partial digestion means cleaving the DNA at only a subset of restriction sites. This can be achieved by adjusting the digestion conditions, for example, concentration of enzymes and time of incubation. Partial digestion is required in cases of cloning a fragment of

Table 1 Characteristics and applications of commonly used enzymes of molecular biology

Name of the enzyme	Characteristic	Application
Lambda exonuclease	This exonuclease, which is purified from *E. coli* cells that have been infected with bacteriophage λ, catalyses the stepwise and processive hydrolysis of duplex DNA from 5'-phosphoryl termini; liberating 5'-mononucleotides. λ exo will not degrade 5'-OH termini.	1. Converting double-stranded DNA into single-stranded DNA that can be subjected to the dideoxy sequencing method. 2. Removing 5'-protruding ends from duplex DNA for tailing with terminal transferase.
Double-stranded 3'→5' Exonucleases or Exonuclease III	Exo III is a multifunctional enzyme that catalyses hydrolysis of several types of phosphodiester bonds in double-stranded DNA.	1. Preparing strand-specific radioactive probes, in conjunction with the Klenow fragment. 2. Preparing single-stranded DNA templates for sequencing by the dideoxy technique. 3. Constructing unidirectional deletions; this property is useful for constructing a set of unidirectional deletions from a given position in cloned DNA, that are used for DNA sequencing without prior restriction site mapping.
Mung bean nuclease enzyme	It is a highly specific single-stranded endonuclease. Reactions are normally performed at pH 5.0. Mung bean nuclease activity is significantly decreased at NaCl concentrations 200 mM.	1. Mung bean nuclease is efficient in cleaving immediately adjacent to the last hybridized base pair without removing any of the base-paired nucleotides. 2. It precisely deletes overhanging bases that result from restriction endonuclease cleavage. 3. The disadvantages of mung bean nuclease are that it is more expensive, and that its activity is more sensitive to reaction conditions.
Micrococcal nuclease	Micrococcal nuclease from *Staphylococcus aureus* is a relatively non-specific nuclease that cleaves single- and double-stranded DNA and RNA to oligo- and mononucleotides with 3'-phosphates. The enzyme is more active on single-stranded nucleic acids. The enzyme is strictly dependent on calcium for activity and hence can be inactivated by Ca^{2+}-specific chelating agents, such as ethylene glycol tetraacetic acid (EGTA).	1. Studying chromatin structure. 2. Removing nucleic acid from crude cell-free extracts without destroying enzyme activities. 3. Digestion of nucleic acid is performed under mild buffer and ionic conditions in the presence of $CaCl_2$. When digestion is complete, micrococcal nuclease is inactivated with EGTA
Deoxyribonuclease I (DNase I)	DNase I from bovine pancreas is an endonuclease that degrades double-stranded DNA to produce 3'-OH	1. Introduces nicks in duplex DNA, which then serve as primer sites to initiate DNA synthesis by *E. coli* DNA polymerase I.

Contd...

Table 1 *Contd...*		
Name of the enzyme	*Characteristic*	*Application*
	oligonucleotides. In the presence of Mg^{2+}, DNase I produces nicks in duplex DNA, while in the presence of Mn^{2+}, the enzyme produces double-stranded breaks in the DNA.	2. Cloning random DNA fragments by catalysing double-stranded cleavage of DNA in the presence of Mn^{2+}.
Ribonuclease A	RNase A from bovine pancreas is a relatively small (~14 kDa) endoribonuclease that specifically hydrolyses RNA after C and U residues. The reaction generates a $2'$: $3'$ cyclic phosphate, which is then hydrolysed to the corresponding 3'-nucleoside phosphate.	1. Mapping and quantitating RNA species using the RNase protection assay in conjunction with RNase T1. 2. Hydrolysing RNA that contaminates DNA preparations. 3. RNA sequencing. 4. Blunt-ending double-stranded cDNA in conjunction with RNaseH.
Ribonuclease I	RNase I from *E. coli* is a periplasmic endoribonuclease with little or no sequence specificity. RNase I shows a marked preference for single-stranded RNA over double-stranded RNA and unlike the RNaseA-type enzymes, it is not inactivated by the placental RNase inhibitor. RNase I is active in the presence of EDTA and is responsible for >90% of the ribonuclease activity of *E. coli* extracts purified in the presence of ethylenediaminetetraacetic acid (EDTA).	1. Analysing RNA structure. 2. Mapping and quantitating RNA species using the RNase protection assay. 3. Degrading ssRNA to mono-, di-, and trinucleotides. 4. Determining protein-binding sites on RNA using RNase footprinting. 5. Removing RNA from DNA preparations.
Ribonuclease T1	RNase T1 from *Aspergillus oryzae* is an endoribonuclease that specifically hydrolyses RNA after G residues. The reaction generates a $2':3'$ cyclic phosphate, which is then hydrolysed to the corresponding $3'$-nucleoside phosphate. RNase T1 is difficult to inactivate and is active under a wide range of reaction conditions, including in the presence of EDTA.	1. Mapping and quantitating RNA species using the RNase protection assay in conjunction with RNaseA. 2. RNA sequencing in conjunction with base-specific RNases CL3 and U2. 3. Analysing RNA structure. 4. Determining the level of RNA transcripts synthesized in vitro from DNA templates containing a "G-less cassette."
Ribonuclease HI	RNase HI (also known as RNaseH) from *E. coli* is an endoribonuclease that specifically hydrolyses the phosphodiester bonds of RNA in RNA–DNA duplexes to generate products with 3'-OH and 5'-phosphate termini	1. Facilitating the synthesis of second-strand cDNA by nicking the mRNA strand of the RNA–DNA duplex produced during first-strand synthesis of cDNA. 2. Creating specific cleavages in RNA molecules by using synthetic deoxyoligonucleotides to create local regions of RNA–DNA duplexes

Contd...

Table 1 *Contd...*

Name of the enzyme	Characteristic	Application
Ribonuclease III	RNase III from *E. coli* is a metal-dependent endoribonuclease specific to double-stranded RNA with a minimum size of two helical turns and is implicated in the maturation of rRNA as well as some tRNAs and mRNAs.	1. Generating siRNA for use in RNAi experiments. 2. In vitro microRNA processing. 3. Analysing RNA structure.
Dicer	Human Dicer belongs to the RNase III superfamily and is a metal-dependent endoribonuclease implicated in the maturation of microRNAs (miRNAs). Like RNase III, Dicer cleaves dsRNA to leave 5′-phosphate and 3-hydroxyl termini with 2-nt 3-overhangs, and the natural substrates of Dicer (pre-microRNAs) have these same overhangs.	1. Generating small interfering RNA (siRNA) for use in RNA interference (RNAi) experiments. 2. In vitro microRNA processing. 3. Analysing RNA structure.
RNA ligases	T4 RNA ligase 1, also known as T4 RNA ligase, is the product of *gene 63* of the T4 phage. It is typically purified as a recombinant enzyme but was originally isolated from phage-infected. T4 RNA ligase 1 catalyses the adenosine triphosphate (ATP)-dependent covalent joining of single-stranded 5′-phosphoryl termini of DNA or RNA to single-stranded 3′-hydroxyl termini of DNA or RNA. T4 RNA ligase 2 also catalyses the joining of a 3′-OH terminus of RNA to a 5′-phosphorylated RNA or DNA; unlike T4 RNA ligase 1, this enzyme prefers double-stranded substrates.	1. Radioactive labelling of 3′-termini of RNA. 2. Circularizing oligodeoxyribonucleotides and oligoribonucleotides that have a 5′-phosphate terminus and a 3′-hydroxyl terminus. 3. Ligating oligomers for oligonucleotide synthesis. 4. One such use is to synthesize oligomers that contain internally labelled oligomers at specific residues. 5. Creating single-stranded RNA/DNA hybrids. 6. Ligating ssDNA oligos to cDNA for construction of cDNA libraries. 7. 5′- and 3′-end mapping and sequencing of mRNA. T4 RNA ligase 1 can be used to determine the length of a 3′-poly(A) tail. 8. Chemical misacylation of tRNA.

Table 2 Enzymes used in molecular cloning

Enzyme type/application	Enzyme	Characteristic
Methylating enzymes	Dam methyltransferase Dcm methyltransferase	Methylation of DNA Methylation of DNA
Restriction endonucleases DNA-dependent DNA polymerase	DNA polymerase I (holoenzyme)	Endonucleases—double strand, specific Synthesizes DNA complementary to a DNA template

Table 2 *Contd...*

Enzyme type/application	Enzyme	Characteristic
	Large fragment of DNA polymerase I (Klenow)	cDNA synthesis of second strand, blunt-end generation by filling in 3'-recessed ends
	Bacteriophage T4 DNA polymerase	Blunt-end generation by removal of 3'-protruding ends, by filling in 3'-recessed ends
	Bacteriophage T7 DNA polymerase	Blunt-end generation by removal of 3'-protruding ends, by filling in 3'-recessed ends
	Thermostable DNA-dependent DNA polymerase	Used in polymerase chain reaction Polymerase chain reaction (PCR)
RNA-dependent DNA polymerase	Reverse transcriptase	cDNA synthesis from RNA
DNA-dependent RNA polymerase	Bacteriophage SP6 RNA polymerase	Transcription, promoter-specific
	Bacteriophage T7 RNA polymerase	Transcription, promoter-specific
	Bacteriophage T3 RNA polymerase	Transcription, promoter-specific
Ligases: DNA /RNA joining enzymes	Bacteriophage T4 DNA ligase	Joins DNA molecules by forming phosphodiester linkages between DNA segments
	E. coli DNA ligase	Ligations of DNA
	Bacteriophage T4 RNA ligase	Ligations of RNA
	Thermostable DNA ligase	Ligations of DNA
Enzymes modifying the ends of DNA	Alkaline phosphatase	Removes phosphate groups from 5'-ends of single-stranded or double-stranded DNA or RNA
	Bacteriophage T4 polynucleotide kinase	Labelling DNA at 5'-ends; polynucleotide
	Terminal transferase	Adds nucleotides to the 3'-ends of DNA or RNA, useful in homopolymer tailing
	Poly(A) polymerase	Poly(A) tailing of RNA
Nucleases	DNase I (pancreatic) endo	Produces single-stranded nicks in DNA endonucleases—double-strand, non-specific
	Bal 31 nuclease	Progressively shortens DNA, degradation of DNA, exonucleases—double-strand, 5'→3' and 3'→5'; endonucleases—single strand
	Nuclease S1	Degrades ssDNA and RNA, Endonucleases—single strand
	Mung bean nuclease	Endonucleases—single strand
	Exonuclease III	Degradation of DNA, exonucleases—double strand, 3'→5'

Contd...

Table 2 *Contd...*

Enzyme type/application	Enzyme	Characteristic
	Exonuclease VII	Exonuclease—single strand (3' and 5')
	Bacteriophage lambda exonuclease	Degradation of DNA, exonucleases—double strand, 5'→3'
	Ribonuclease H	Cleaves and digests RNA strand of RNA–DNA duplex; degradation of RNA in RNA–DNA duplexes
	Ribonuclease A (pancreatic)	Cleaves and digests RNA (not DNA); Degradation of RNA
	Ribonuclease T1	Degradation of RNA
Proteolytic enzymes	Proteinase K	Digestion of proteins
	Lysozymes	Lysis of cell wall

DNA in which the site(s) used for cloning is also present internally, as well as for restriction mapping.

Star Activity

Sometimes restriction enzymes recognize and cleave the DNA strand at the recognition site with asymmetrical palindromic sequence, for example, *Bam*HI cuts at the sequence GA^TCC, but under extreme conditions, such as low ionic strength, it will cleave in any of the following sequence NGA^TCC, GPOA^TCC, GGN^TCC. Such an activity of the restriction endonuclease is called star activity.

Enzyme Quality Assessment

For any cloning experiment, restriction digestion is the first step. Restriction enzymes in a molecular biology laboratory are usually available from many different commercial sources. For successful digestion, the enzymes should be free from contaminants like exonucleases, endonucleases, and phosphatases. Contaminating exonucleases can chew away the cohesive ends, thereby reducing the production of recombinants. Contaminating phosphatases can remove the terminal phosphate residues, thereby preventing ligation.

A typical quality check can be done as follows. The restriction endonuclease is used to digest DNA fragments up to excessive overdigestion. The fragments generated are ligated and recut with the same restriction endonuclease. Ligation can occur only if the 3'- and 5'-termini are left intact. Only those molecules that perfectly restore the recognition site will be re-cleaved. A normal banding pattern after cleavage indicates that both the 3'- and 5'-termini are intact and the enzyme preparation is free of detectable exonucleases and phosphatases.

SUMMARY

- The major tools for genetic engineering are the enzymes that catalyse specific reactions on DNA molecules.

- DNA polymerase I from *E. coli*, possessing 5'→3' DNA polymerase activity, 3'→5' exonuclease activity, and 5'→3' exonuclease activity, is extensively used for recombinant DNA technology. DNA Pol I has been engineered to create the large C-terminal fragment or Klenow fragment, which lacks the 5'→3' exonuclease activity and has wider applications in molecular biology. Phage-encoded DNA polymerases, such as T7 and T4 DNA polymerases, find specific uses in recombinant DNA technology.

- Thermostable DNA polymerases isolated from extremophiles are used in DNA amplification by polymerase chain reaction (PCR). Terminal deoxynucleotidyl transferase (TdT) is a template-independent DNA polymerase that catalyses the addition of deoxynucleotides to the 3'-hydroxyl terminus of DNA.

- Conversion of RNA to DNA is mediated by reverse transcriptase obtained from retroviruses and is useful for the construction of cDNA libraries.

- Alkaline phosphatase catalyses the hydrolysis of 5'-phosphate residues from DNA, RNA, and ribo- and deoxyribonucleoside triphosphates. The process of removing phosphate groups is called dephosphorylation, very often required for successful cloning. The enzyme polynucleotide kinase catalyses the transfer of a phosphate group from ATP to the 5'-ends of a polynucleotide DNA or RNA.

- Single-strand-specific exonucleases and endonucleases, which chew up single-stranded DNA molecules, can be used for the creation of DNA molecules of different sizes.

REVISION QUESTIONS

1. Illustrate with figure the function of the following enzymes: DNA polymerase, Klenow fragment, DNA ligase, polynucleotide kinase, alkaline phosphatase, terminal transferase.

2. Distinguish between exonuclease and endonuclease.

3. How does reverse transcriptase help in creating DNA library?

4. Suppose a particular restriction enzyme has three sites in (a) a linear DNA fragment and (b) a circular DNA. How many fragments do you expect to be produced in each case?

5. Suppose you like to carry out a restriction enzyme digestion reaction. What ingredients will you require from (a) water, (b) ethyl alcohol, (c) DNA, (d) buffer, (e) alkaline phosphatase, (f) restriction enzyme, and (g) DNA ligase.

19

Agarose Gel and Polyacrylamide Gel Electrophoresis

OBJECTIVES

After reading this chapter, the student will be able to:

- Explain agarose gel electrophoresis
- Discuss the factors affecting the rate of migration of DNA through agarose gel
- Describe electrophoresis buffer
- Explain pulsed field gel electrophoresis
- Understand polyacrylamide gel electrophoresis for small DNA molecules

INTRODUCTION

Fragments of DNA can be separated by electrophoresis through an agarose gel (or for very small fragments, a polyacrylamide gel). Electrophoresis is the technique of separation of charged molecules under the influence of an electrical field in which the charged molecules migrate in the direction of the electrode bearing the opposite charge, namely, cationic (positively charged) molecules moving towards the cathode (negative electrode) and anionic (negatively charged) molecules travelling towards the anode (positive electrode). Electrophoresis can be carried out in free solutions, for example, capillary electrophoresis, or in a stabilizing support material like thin layer plates, films, and gels. Agarose or polyacrylamide provides the support matrix for DNA.

Agarose and polyacrylamide gel electrophoresis are presently the state-of-the-art techniques in the field of molecular biology and are routinely used to separate, identify, and purify DNA fragments. The technique is simple and fast in performance. It has largely replaced the earlier density gradient centrifugation technique. The location of DNA within the gel can be easily detected by directly staining the gel with low concentration of fluorescent intercalating dye such as ethidium bromide or SYBR Gold.

PRINCIPLE

The DNA molecule contains high negative charge due to the phosphates that form the sugar-phosphate backbone of each DNA strand. A small DNA fragment has less negative charge than a large DNA fragment, as fewer phosphates are present in it. The overall charge per unit length for both small and large DNA molecules is, however, identical. If an electric field is applied to a sample containing small and large DNA fragments in free solution, both the molecules will move to the positive electrode at the same rate, assuming that friction is negligible in free solution. Thus it will not lead to the separation of DNA molecules. So a mechanism by which DNA molecules can be separated is to increase the amount of friction so that small DNA molecules move to the anode faster by virtue of having less friction than larger DNA molecules. Running the DNA fragments through a gel can provide the necessary friction to separate DNA fragments of different sizes.

The electric field strength depends on the length of the gel matrix and the potential difference at the ends (voltage applied). The migration velocity of DNA depends on the frictional force imposed by the gel matrix and the type of the matrix and its concentration.

AGAROSE GEL ELECTROPHORESIS

Agarose is a linear polymer of alternating residues of D- and L-galactose joined by alpha (1 to 3) and beta (1 to 4) glycosidic linkage. The L-galactose residue has an anhydro bridge between the three and six positions (Figure 1). Agarose forms helical fibres, which aggregate into supercoiled structures. Upon gelation, a three-dimensional mesh is generated through which DNA can pass through. Most agarose gels are made ranging between 0.7% and 2%. A 0.7% gel shows good separation (resolution) of large

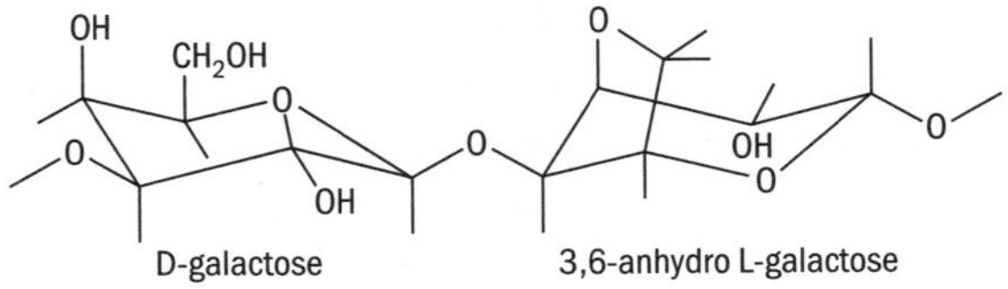

Figure 1 Agarose

DNA fragments (5–10 kb), and a 2% gel shows good resolution for small fragments (0.2–1 kb).

FACTORS AFFECTING THE RATE OF MIGRATION OF DNA THROUGH AGAROSE GEL

The following factors determine the rate of migration of DNA through agarose gels.

❖ *The molecular size of the DNA* Larger molecules migrate more slowly compared to smaller fragments as these molecules face greater frictional drag while moving through the pores of the gel. A rough estimate suggests that the rate of migration of a double-stranded DNA molecule varies inversely to the $\log_{10}$ of the number of base pairs.

❖ *The concentration of agarose* The migration of DNA fragment of a given size varies with the agarose concentration. It is expected because the pores of the agarose matrix depend on its concentration. There is a linear relationship between the logarithm of the electrophoresis mobility of the DNA (μ) and the gel concentration (c), which can be expressed as

$$\log\mu = \log\mu_0 - K_r.c$$

where μ is the free electrophoresis mobility of DNA and K_r is the retardation coefficient, a constant related to the properties of the gel and the size and shape of the migrating molecules.

❖ *The conformation of the DNA* DNA exists in different conformations, which can be separated in agarose gel electrophoresis. The superhelical circular, nicked circular, and linear forms of DNA migrate through agarose gels at different rates. The relative mobilities of the three forms depend on the concentration and type of agarose, the strength of the applied current, the ionic strength of the buffer, and the density of superhelical twists in the superhelical circular form of DNA.

❖ *The presence of ethidium bromide in the gel and electrophoresis buffer* Ethidium bromide is a fluorescent dye that intercalates between bases of nucleic acids. It can be incorporated into agarose gels or added to samples of DNA before loading to enable visualization of the fragments within the gel. As might be expected, binding of ethidium bromide to DNA alters its mass and rigidity and, therefore, its mobility.

❖ *The applied voltage* At low voltages, the rate of migration of linear DNA fragments is proportional to the voltage applied. However, as the strength of the electric field is raised, the mobility of high-molecular-weight fragments increases differentially. Thus the effective range of separation in agarose gels decreases as the voltage is increased. To obtain maximum resolution of DNA fragments >2 kb in size, agarose gels should be run at no more than 5–8 V/cm.

❖ *The type of agarose* The two major classes of agarose are standard agaroses and low-melting-temperature agaroses. A third and growing class consists of intermediate-melting/gelling-temperature agaroses, exhibiting properties of each of the two major classes. Within each class are various types of agaroses that are used for specialized applications.

ELECTROPHORESIS BUFFER

The electrophoretic mobility of DNA is affected by the composition and ionic strength of the electrophoresis buffer. In the absence of ions, for example, if one mistakenly uses water instead of buffer, electrical conductivity is minimal and DNA migrates slowly, if at all. In a buffer of high ionic strength, electrical conductance is very efficient, and significant amount of heat is generated, even when moderate voltages are applied. In the worst case, the gel melts and the DNA denatures. The most commonly used buffers for duplex DNA are TAE, or Tris-acetate-ethylenediaminetetraacetic acid (EDTA), and TBE (Tris-borate-EDTA). Buffers not only establish a pH, but also provide ions to support conductivity.

Equipment and Supplies

The equipment and supplies necessary for conducting agarose gel electrophoresis are relatively simple and include the following.

❖ Horizontal gel electrophoresis apparatus or electrophoresis chamber.
❖ Gel casting platform or trays (usually of UV-transparent plastic and available in a variety of sizes). The open ends of the trays are closed with tape while the gel is being cast, which is removed prior to electrophoresis.
❖ Sample combs to form sample wells in the gel.
❖ DC power supply.
❖ Transilluminator (a UV lightbox) or gel-documentation system.
❖ Electrophoresis grade agarose.
❖ Electrophoresis buffer, usually TAE or TBE.
❖ Loading buffer.
❖ Ethidium bromide.

Procedure

To pour a gel, agarose powder is mixed with electrophoresis buffer to the desired concentration and then heated until completely melted. Ethidium bromide is often added to the solution (final concentration 0.5 µg/ml) at this stage to visualize DNA after electrophoresis. The agarose solution is then cooled to 60°C and poured onto the casting tray containing a suitable comb. It is allowed to solidify at room temperature.

After solidification of the gel, the comb is removed with caution so that the bottom of the well is intact. The gel is inserted horizontally into the electrophoresis chamber

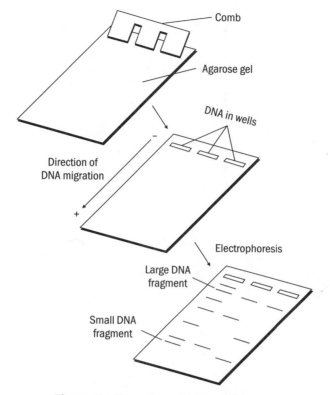

Figure 2   Agarose gel electrophoresis

and just covered with buffer. DNA samples are then mixed with loading buffer and pipetted into the sample wells, the lid and power leads are placed on the apparatus, and current is applied (Figure 2). DNA will migrate towards the positive electrode, which is usually coloured red.

The tracking dyes are added along with the loading buffer to visualize the migration of DNA in the gel. Two tracking dyes of different molecular weights are used—one moves faster and is called the leading dye and the other moves slower and is called the lagging dye. The distance through which the DNA has migrated in the gel can be judged by visually monitoring the migration of the tracking dye. Bromophenol blue and xylene cyanol are the two dyes that migrate through agarose gels at roughly the same rate as double-stranded DNA fragments of 300 and 4000 base pairs, respectively (Plate 1).

After adequate migration of the DNA fragments, the gel is removed from the apparatus and visualized either in a transilluminator under UV light or in a gel-documentation system, which also operates in UV but has a CCD camera to take photograph, which is visible on a computer screen. Ethidium bromide may be added during gel casting, or the gel may be soaked in a dilute solution of ethidium bromide after electrophoretic run. Ethidium bromide is a known mutagen and should be treated as a hazardous chemical and handled with gloves.

Composition of Reagents Required

* ❖ Loading buffer
* ❖ Glycerol loading dye (6X)
* ❖ Bromophenol blue (0.26 g)
* ❖ Glycerol (30 ml)
* ❖ 100 ml with nanopure water

Usually xylene cyanol is also used with bromophenol blue in equal proportion. The dyes are added for visual monitoring of how far the electrophoresis has proceeded. Xylene dye moves much slower than bromophenol blue and is helpful for monitoring long runs.

If DNA fragments digested with restriction enzymes are used for electrophoresis, 0.1 M EDTA (stops restriction enzyme digestion) and 1% SDS (denatures the restriction enzyme to stop the reaction) are used.

* ❖ Running buffer
* ❖ TAE (Tris-acetate-EDTA) electrophoresis buffer
* ❖ 50× stock solution
* ❖ 242 g Tris base
* ❖ 57.1 ml glacial acetic acid
* ❖ 37.2 g $Na_2EDTA \cdot 2H_2O$
* ❖ H_2O to 1 L
* ❖ Working solution, pH ~ 8.5
* ❖ 40 mM Tris-acetate
* ❖ 2 mM $Na_2EDTA \cdot 2H_2O$
* ❖ TBE (Tris-borate-EDTA) electrophoresis buffer
* ❖ 10× stock solution, 1 L
* ❖ 108 g Tris base (890 mM)
* ❖ 55 g boric acid (890 mM)
* ❖ 40 ml 0.5 M EDTA, pH 8.0

In case of TBE, borate is the ion, which allows the generation of an electric field in the gel set-up, while for TAE, it is acetate. But Tris-borate has a significantly better buffering capacity, which means gel running can be done at high voltages. However, borate also forms complexes with the agarose sugar monomers/polymers, so it is avoided for purification of DNA from gels.

❖ Ethidium bromide solution
❖ 1000× stock solution, 0.5 mg/ml
❖ 50 mg ethidium bromide
❖ 100 ml H$_2$O
❖ Working solution, 0.5 g/ml
❖ Stock diluted by 1:1000 for gels or stain solution
❖ Kept in dark

Agarose

Commercial preparation of agarose varies from batch to batch, and the preparation is not always homogenous. Lower grades of agarose may be contaminated with other polysaccharides, as well as salts and proteins. This variability can affect the gelling/melting temperature of agarose solutions, the sieving of DNA, and the ability of the DNA recovered from the gel to serve as a substrate in enzymatic reactions. These potential problems are minimized if special grades of agarose are used, which are screened for the presence of inhibitors and nucleases and for minimal background fluorescence after staining with ethidium bromide.

The advantages are that the gel is easily poured and is inert to samples, that is, does not denature the samples. The samples can also be recovered. The disadvantages are that gels can melt during electrophoresis, the buffer can become exhausted, and different conformations of DNA may run in unpredictable forms. After the experiment is finished, the resulting gel can be stored in a plastic bag in a refrigerator.

Applications

Agarose gel electrophoresis is used in almost all types of DNA manipulation; a few of its uses include the estimation of the size of DNA molecules following restriction enzyme digestion, analysis of polymerase chain reaction (PCR) products, molecular genetic diagnostic or genetic fingerprinting.

PULSED FIELD GEL ELECTROPHORESIS

In DNA gel electrophoresis, negatively charged DNA molecules are pulled through a gel matrix by an electric field. An agarose gel possesses a set of pores of varying sizes. The small molecules pass easily through most of the pores, while the larger molecules have to squeeze through small pores. The mobility of a molecule is related to the fraction of pores that it can easily move through, a process known as "sieving". DNA molecules of size larger than a certain size cannot be sieved and, therefore, do not even enter the gel. Depending on the percentage of agarose and the condition of the run, the size of DNA that can pass through may vary, but fragments of size 20–40 kb migrate with limited mobility. Schwartz and Cantor in 1984 described pulsed field gel electrophoresis (PFGE), introducing a new way to separate large DNA molecules by cyclically varying the orientation of the electric field in the gel during the run. There was then a succession of papers describing improved instrumentation for PFGE. PFGE permits cloning and analysis of a smaller number of very large pieces of a genome.

Theory

During continuous field electrophoresis, DNA above 30–50 kb migrates with the same mobility regardless of size. This is seen in a gel as a single large diffuse band. If, however, the field direction is periodically changed, the various lengths of DNA react to the change at differing rates, that is, larger pieces of DNA are slower to realign their charge when field direction is changed, while smaller pieces are quicker. Over the course of time, with the consistent changing of directions, each band begins to separate more and more even at very large lengths. Thus separation of very large DNA pieces using PFGE is possible (Figure 3).

Procedure

The procedure for this technique is relatively similar to performing a standard gel electrophoresis except that the voltage is periodically switched among three directions—one that runs through the central axis of the gel and two that runs at an angle of 120° on either side. The pulse times are equal for each direction, resulting in a net forward

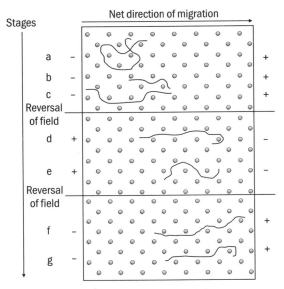

Figure 3 PFGE principle

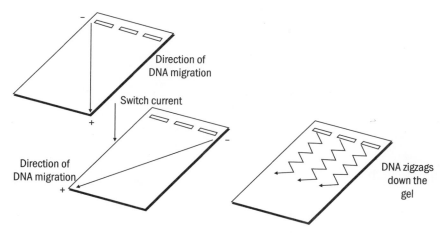

Figure 4   Pulsed field gel electrophoresis

migration of the DNA. For extremely large bands (up to around 2 Mb), switching-interval ramps can be used that increase the pulse time for each direction over the course of a number of hours, for instance, increasing the pulse linearly from 10 s at 0 h to 60 s at 18 h (Figure 4).

This procedure takes much longer time for resolution of fragments than normal gel.

Applications

Applications of PFGE are numerous and diverse. These include cloning of large plant DNA using yeast artificial chromosomes (YACs) and P1 cloning vectors; identification of restriction fragment length polymorphisms (RFLPs) and construction of physical maps detecting in vivo chromosome breakage and degradation; and determining the number and size of chromosomes ("electrophoretic karyotype") from yeasts, fungi, and parasites such as *Leishmania, Plasmodium*, and *Trypanosoma*, genotyping and genetic fingerprinting, in epidemiological studies of pathogenic bacteria.

POLYACRYLAMIDE GEL ELECTROPHORESIS FOR SMALL DNA MOLECULES

Cross-linked chain of polyacrylamide has been used as a matrix for electrophoresis in cases of separation of smaller DNA fragments, which cannot be separated by agarose gel. The cross-linked polyacrylamide has a much lower pore size to allow DNA fragments of smaller size. Single-stranded DNA can also be separated according to size and conformation (Figure 5).

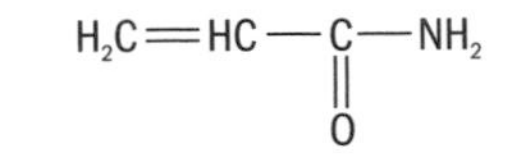

H₂C=HC—C—NH₂
||
O

Acrylamide

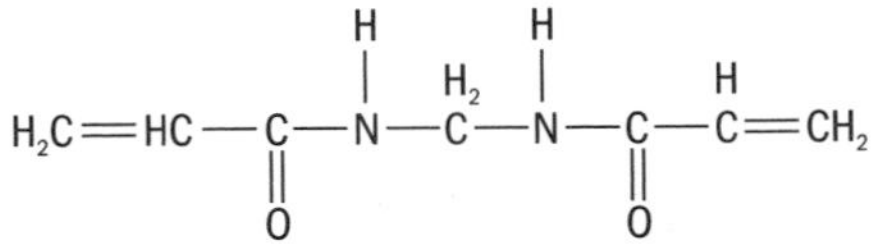

N,N'-methylenebisacrylamide

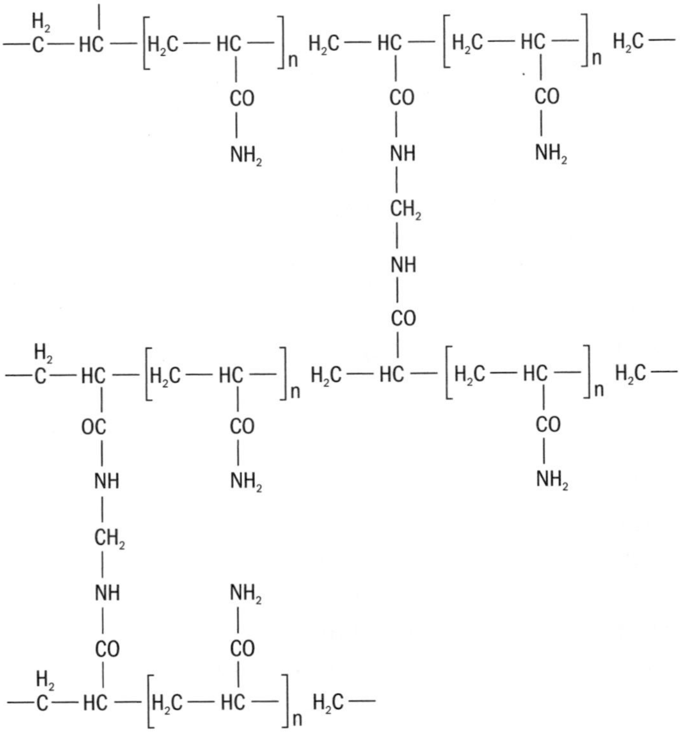

Cross-linked polyacrylamide

Figure 5 Polyacrylamide

In the presence of free radicals, the monomers of acrylamide are polymerized into long chains. These free radicals are generated by ammonium persulphate by TEMED, or N,N,N'N'-tetramethylene diamine. When mixed with bifunctional cross-linking agent N,N'-methylenebis acrylamide, the copolymerization reaction induces three-dimensional ribbon-like cross-linked structure, and the size of the pores depends on the concentration of the acrylamide monomer as well as on the bisacrylamide or more precisely on their ratio.

Polyacrylamide gels have the following three major advantages over agarose gels.
1. High resolving power.
2. Can accommodate more DNA without the loss of resolution.
3. DNA recovered from polyacrylamide gels is extremely pure and can be used in a downstream experiment without further purification.

The following are the two types of polyacrylamide gels commonly used.
1. Denaturing polyacrylamide gels, polymerized in the presence of usually urea, which suppresses base pairing in nucleic acids. Denatured DNA migrates through these gels at a rate that is almost completely independent of its base composition and sequence. This type of gel is used for purification and separation of single-stranded DNA. More specifically, this type of gel is used in DNA sequencing, for isolation of radiolabelled DNA probes, and in analysis of the products of nuclease SI digestions.
2. Non-denaturing polyacrylamide gels are used for the separation and purification of fragments of double-stranded DNA. Non-denaturing polyacrylamide gels are used chiefly to prepare highly purified fragments of DNA (this protocol) and to detect protein–DNA complexes structure of polyacrylamide.

Buffers and Solutions for Polyacrylamide Gel Electrophoresis

Ammonium persulphate (10% w/v) is used as a catalyst for the copolymerization of acrylamide and bisacrylamide gels. The polymerization reaction is driven by free radicals that are generated by an oxido-reduction reaction in which a diamine (for example, TEMED) is used as the adjunct catalyst. The ratio of acrylamide to bisacrylamide is 29:1 (30% w/v).

5× TBE Electrophoresis Buffer

Polyacrylamide gels are poured and run in 0.5× or 1× TBE at low voltage (1–8 V/cm) to prevent denaturation of small fragments of DNA by heating. Other electrophoresis buffers such as 1× TAE can be used, but they are not as good as TBE. The gel must be run more slowly in 1× TAE, which does not provide as much buffering capacity as TBE. For electrophoresis runs greater than 8 h, it is recommended that 1× TBE buffer be used to ensure that adequate buffering capacity is available throughout the run.

SUMMARY

- Agarose gel electrophoresis is used to separate, identify, and purify DNA fragments. DNA molecules are negatively charged and are allowed to move through an agarose matrix under the influence of an electric field. Shorter molecules migrate more easily through the pores of the matrix, and DNA fragments are separated based on their size. The migration of DNA molecules through the gel also depends on the concentration and type of agarose, conformation of DNA, applied voltage, and the type of buffer.

- The pulsed field gel electrophoresis (PFGE) separates large DNA molecules by cyclically varying the orientation of the electric field in the gel during the run. PFGE permits cloning and analysis of a smaller number of very large pieces of a genome.

- The separation and analysis of small DNA fragments are done by using polyacrylamide as the matrix instead of agarose.

- Following agarose and polyacrylamide gel electrophoresis, the DNA molecules are visualized by staining with ethidium bromide or SYBR Gold.

REVISION QUESTIONS

1. Use a drawing to illustrate the principle of DNA gel electrophoresis. Indicate roughly the comparative electrophoretic mobilities of DNAs with 200 base pairs, 500 base pairs, and 1000 base pairs.

2. In gel electrophoresis, DNA molecules migrate according to size. An intact plasmid DNA produces three bands, while the plasmid DNA digested with a restriction enzyme that cuts the DNA once migrate with a single band. Explain.

3. In gel electrophoresis, why is the marker DNA (for example, lambda cut with *Hind*III) useful?

Detection and Extraction of DNA from Gels

INTRODUCTION

Nucleic acids subjected to agarose gel electrophoresis can be detected by staining. The most commonly used stains to visualize DNA are ethidium bromide and SYBR Gold. Polyacrylamide gels are also stained by these dyes.

STAINING DNA GELS USING ETHIDIUM BROMIDE

Ethidium bromide (3,8-diamino-6-ethyl-5-phenylphenanthridium bromide) is commonly used for direct visualization of DNA in gels (Figure 1). The dye intercalates between the stacked bases of nucleic acids and fluoresces in the red-orange range (560 nm) when illuminated with UV light (260–360 nm). The dye is sensitive enough to detect less than 5 ng of DNA. The main advantage of the use of EtBr is that it can bind to the DNA with little or no sequence specificity.

EtBr can be used to visualize both single- and double-stranded DNA. However, the affinity of the dye for single-stranded nucleic acids is relatively low, and the fluorescent yield is poor. In fact, the little fluorescence observed is attributable to the small intra-strand duplexes formed. Moreover, closed circular DNA binds less drugs at saturation compared to nicked open circular form of DNA.

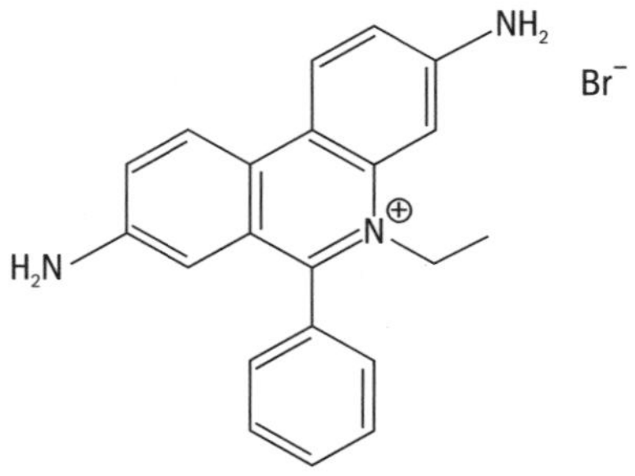

Figure 1 Ethidium bromide

The reaction between DNA and EtBr is reversible, but the dissociation reaction is very slow. In practice, dissociation is achieved by passing the complex through cation exchange resin column.

Generally, EtBr is added to the gel for convenience and the bands visualized under UV just after the run. But for accurate size determination of DNA bands, it is advised to stain the gels after the run. The gel is soaked in EtBr solution for 30–45 min for staining. Destaining is usually not required. However, in case of very small quantity of DNA (<10 ng), detection is made easier if the background fluorescence is reduced by soaking the stained gel in water or 1 mM $MgSO_4$ for 20 min in room temperature.

DETECTION OF DNA IN GELS CONTAINING SYBR GOLD

The SYBR Gold nucleic acid gel stain is more sensitive compared to EtBr, and it can detect double- or single-stranded DNA or RNA in electrophoretic gels, using standard UV transilluminators. It is an unsymmetrical cyanine dye (Figure 2), which has a low intrinsic fluorescence but shows 1000 times enhancement of fluorescence when bound to nucleic acids. The dye has two excitation maxima, one at 495 nm in the visible range and the other at 300 nm in the UV range. The emission is at 537 nm. The stain penetrates the agarose gel rapidly and does not require destaining due to the low intrinsic fluorescence of the unbound dye. The presence of the dye in stained gels at

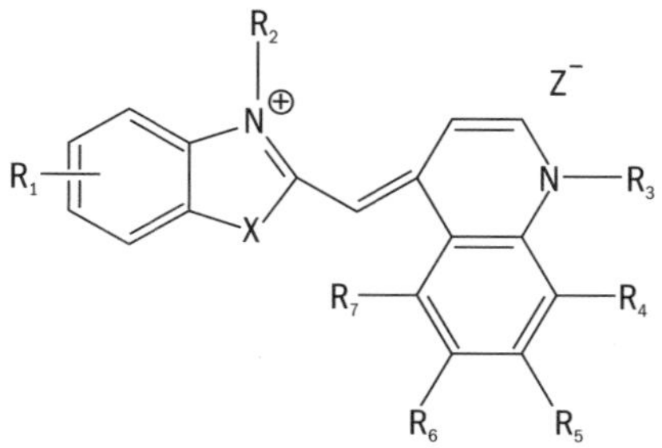

Figure 2 SYBR Gold

standard staining concentrations does not interfere with restriction endonucleases, T4 DNA ligase, Taq polymerase, or with Southern or Northern blotting. It can be easily removed by ethanol precipitation of nucleic acids and can be used for downstream experiments thereafter. The nucleic acids stained with SYBR Gold can be detected in gels using laser scanners or standard UV transilluminators. When excited by standard transillumination at 300 nm, it gives bright gold fluorescent signals that can be captured by conventional black and white Polaroid films or CCD-based image detection system. The high cost of the dye precludes its use from routine staining of gels. However, it can be cost effective as an alternative to using radiolabelled DNA in techniques like SSCP and DGGE.

SYBR Gold is used to stain the DNA by soaking it after electrophoresis in a 1:10000 fold dilution of the stock dye solution. It should not be added before the separation is over, as its presence in the hardened gel will cause severe distortions in the electrophoretic mobility of the DNA or RNA. The dye is sensitive to fluorescent light, and working solutions of the dye should be freshly made in electrophoresis buffer and stored at room temperature.

PHOTOGRAPHY OF DNA IN GELS

Photography of EtBr-stained gels can be made using transmitted or incident UV light. Most transilluminators emit UV at a wavelength of 302 nm. In the more advanced forms, gel images are directly transmitted to a computer and viewed in real time. The image can be manipulated with respect to the field, focus, and cumulative exposure time prior to printing. Individual images can be printed, saved, and stored electronically in several file formats and further manipulated using several image analysis software programs. Images can also be visualized by using integrated systems containing light systems, thermal printers, and so on, or in a gel documentation system with CCD camera. A typical polyacrylamide gel picture in a gel documentation system is shown in Figure 3.

EXTRACTION OF DNA FROM GELS

Although many methods have been developed throughout these many years, most have not proven to be efficient as there are various problems in the recovery of DNA from agarose. They are as follows.

❖ *DNA recovered from agarose gels is frequently difficult to ligate, digest, or radiolabel which the charged polysaccharide inhibitors in the eluted DNA cause* Although there has been improvement in the agarose gels, but still, in many cases, DNA is recovered in nonreactive form.

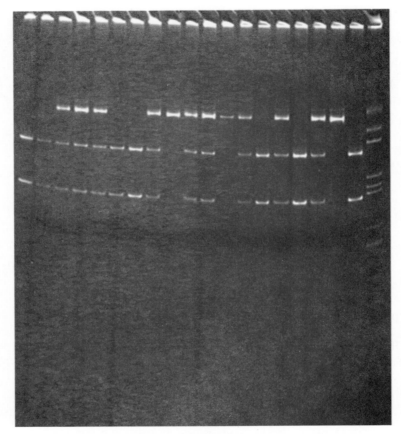

Figure 3 Photography of the DNA in gels

❖ *Inefficient recovery of large fragments of DNA* The efficiency with which DNA is recovered from the gel is a function of its molecular weight. Although most methods give reasonable yields of DNA fragments (<5 kb) in length, none of them are entirely satisfactory for recovery of larger fragments. Larger fragments bind more tenaciously to the solid supports and are more difficult to elute.

❖ *Inefficient recovery of small amounts of DNA* The smaller the amount of DNA in the band, the lower the yield of purified DNA. In certain methods, the loss is so great that it is not worthwhile to attempt to recover bands that contain <500 ng of DNA.

❖ *Inability to recover a number of small fragments simultaneously* Several techniques are labour intensive and consist of individual manipulations. The number of fragments that can be processed at any one time is thus limited.

EXTRACTION OF DNA (DNA RECOVERY)

Following gel electrophoresis, the DNA needs to be eluted for downstream manipulation. Various methods are available for extraction of DNA samples from gels. These include electroelution, use of low melting agarose when DNA is recovered following heating and purified with phenol, digestion with agarase, or via silica membrane spin columns.

Recovery by Electroelution

The DNA fragments are separated by electrophoresis on either a standard or a low melting point agarose gel until the DNA fragment of interest is resolved. Either TAE or TBE electrophoresis buffer can be applied. The portion of the gel containing the restriction fragment to be purified is then physically removed from the remainder of the gel by using a razor blade or scalpel. This agarose slice is placed into a buffer-filled piece of dialysis tubing and again subjected to electrophoresis when the restriction fragment migrates out of the gel slice, which may be further purified or concentrated by column.

Recovery Using Silica Membrane Spin Columns

This approach is rapid. A gel slice containing DNA fractionated through an agarose gel is melted and passed through a silica membrane column in the presence of high salt. Under these conditions, DNA is adsorbed onto the silica membrane. The gel contaminants are subsequently washed away, and DNA is eluted in low-salt buffer. This works well with standard agarose gels, as also for low melting agarose. Although it is more rapid than standard electroelution or organic extraction methods, this approach may result in somewhat lower yields. Silica membrane spin columns for purification of nucleic acids from agarose gels are available from many companies (including Qiagen, Promega, Invitrogen, and Novagen) as kits that include silica membrane spin columns and all appropriate buffers necessary for DNA purification. The nucleic acid is bound to a high quality chromatographic silica matrix in the presence of chaotropic salt. Impurities such as agarose, protein, salt, and dye are removed with wash buffer. The DNA is eluted in water or TE, ready to use in subsequent reactions, for example, ligation, enzymatic restriction, labelling, or sequencing. Silica particles are uniform in size and have smooth surfaces ensuring recovery of DNA fragments up to 100 kb. A typical procedure is illustrated by the flow chart presented in Figure 4.

Application

Gel purification of DNA is a convenient and less time-consuming way to isolate high quality DNA.

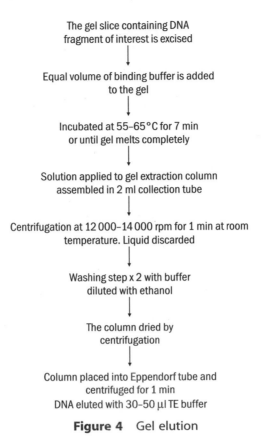

The gel slice containing DNA
fragment of interest is excised

↓

Equal volume of binding buffer is added
to the gel

↓

Incubated at 55–65°C for 7 min
or until gel melts completely

↓

Solution applied to gel extraction column
assembled in 2 ml collection tube

↓

Centrifugation at 12 000–14 000 rpm for 1 min at room
temperature. Liquid discarded

↓

Washing step x 2 with buffer
diluted with ethanol

↓

The column dried by
centrifugation

↓

Column placed into Eppendorf tube and
centrifuged for 1 min
DNA eluted with 30–50 µl TE buffer

Figure 4   Gel elution

SUMMARY

- The DNA fragments need to be eluted from agarose or polyacrylamide gels for downstream manipulation. The methods for extraction of DNA samples from gels include electroelution. Low melting agarose is often used when DNA is recovered following heating.

- The eluted DNA fragments are purified with phenol and digested with agarase to remove agarose or via silica membrane spin columns.

REVISION QUESTIONS

1. How ethidium bromide detects DNA in a gel? What precautions would you take while using ethidium bromide?

2. What advantages does SYBR Gold have over ethidium bromide?

3. Why is the column method of gel extraction preferred to the electroelution method?

Bibliography

Alberts, B. 2003. DNA replication and recombination. *Nature* 421: 431–435

Allison, L. A. 2006. Transcription in eukaryotes. In *Fundamental Molecular Biology*. UK: Blackwell Publishing

Amann, E., J., Brosius, and M., Ptashne. 1983. Vectors bearing a hybrid *trp*–lac promoter useful for regulated expression of cloned genes in *Escherichia coli*. *Gene* 25: 167–178

Amann, E., B., Ochs, and K.-J., Abel. 1988. Tightly regulated tacpromoter vectors useful for the expression of unfused and fused proteins in Escherichia coli. *Gene* 69: 301–315

Arnheim, N., and H., Erlich. 1992. Polymerase chain reaction strategy. *Annual Review of Biochemistry* 61: 131–156

Attardi, G. 1967. The mechanism of protein synthesis. *Annual Review of Microbiology* 21: 383–416

Ausubel, F. M., R., Brent, R. E., Kingston, D. D., Moore, J. G., Seidman, J. A., Smith, and K., Struhl. 2002. *Current Protocols in Molecular Biology*. New York: Wiley

Ballantyne, J., G., Sensabaugh, and J., Witkowski. 1989. *DNA Technology and Forensic Science*. New York: Cold Spring Harbour Laboratory Press

Baltimore, D. 1970. RNA-dependent DNA polymerase in virions of RNA tumour viruses. *Nature* 226: 1209–1211

Banach-Orlowska, M., I. J., Fijalkowska, and R. M., Schaaper. 2005. DNA polymerase II as a fidelity factor in chromosomal DNA synthesis in *Escherichia coli*. *Molecular Microbiology* 58: 61–70

Bansal, M. 2003. DNA structure: revisiting the Watson–Crick double helix. *Current Science* 85: 11

Barnes, W. M. 1992. The fidelity of Taq polymerase catalysing PCR is improved by an N-terminal deletion. *Gene* 112: 29–35

Barnes, W. M. 1994. PCR amplification of up to 35 kb DNA with high fidelity and high yield from λ bacteriophage templates. *Proceedings of the National Academy of Sciences, USA* 91: 2216–2220

Bassett, A., S., Cooper, C., Wu, and A., Travers. 2009. The folding and unfolding of eukaryotic chromatin. *Current Opinion in Genetics and Development* 19: 159–165

Bebenek, K., and T. A., Kunkel. 1989. The use of native T7 DNA polymerase for site-directed mutagenesis. *Nucleic Acids Research* 17: 5408

Benham, C. J. 1985. Theoretical analysis of conformational equilibria in superhelical DNA. *Annual Review of Biophysics and Biophysical Chemistry* 14: 23–45

Benham, C. J., and S. P., Mielke. 2005. DNA mechanics. *Annual Review of Biomedical Engineering* 7: 21–53

Birnboim, H. C. 1983. A rapid alkaline extraction method for the isolation of plasmid DNA. *Methods in Enzymology* 100: 243–255

Birnboim, H. C., and J., Doly. 1979. A rapid alkaline extraction procedure for screening recombinant plasmid DNA. *Nucleic Acids Research* 7: 1513–1523

Blumenthal, T. 1998. Gene clusters and polycistronic transcription in eukaryotes. *Bioessays* 20: 480–487

Blumenthal, T. 2004. Operons in eukaryotes. *Briefings in Functional Genomic Proteomic* 3: 199–211

Bolivar, F., R. L., Rodriguez, P. J., Greene, M. C., Betlach, H. L., Heyneker, H. W., Boyer, J. H., Crosa, and S., Falkow. 1977. Construction and characterization of new cloning vehicles. II. A multipurpose cloning system. *Gene* 2: 95–113

Brummelkamp, T., R., Bernards, and R., Agami. 2002. A system for stable expression of short interfering RNAs in mammalian cells. *Science* 296(5567): 550–553

Calvin, N. M., and P. C., Hanawalt. 1988. High-efficiency transformation of bacterial cells by electroporation. *Journal of Bacteriology* 170: 2796–2801

Carle, G. F., and M. V., Olson. 1984. Separation of chromosomal DNA molecules from yeast by orthogonal field alternation gel electrophoresis. *Nucleic Acids Research* 12: 5647–5664

Carle, G. F., M., Frank, and M. V., Olson. 1986. Electrophoretic separations of large DNA molecules by periodic inversion of the electric field. *Science* 232: 65–68

Challberg, M. D., and P. T., Englund. 1980. Specific labelling of 3′ termini with T4 DNA polymerase. *Methods in Enzymology* 65: 39–43

Chang, L. M., and F. J., Bollum. 1986. Molecular biology of terminal transferase. *CRC Critical Reviews in Biochemistry* 21: 27–52

Chiari, M., and P. G., Righetti. 1995. New types of separation matrices for electrophoresis. *Electrophoresis* 16: 1815–1829

Chien, A., D. B., Edgar, and J. M., Trela. 1976. Thermostable enzymes and fidelity: deoxyribonucleic acid polymerase from the extreme thermophile *Thermusaquaticus*. *Journal of Bacteriology* 127: 1550–1557

Chu, G., 1990. Pulsed-field gel electrophoresis: theory and practice. *Methods* 1: 129–142

Chung, C. T., S. L., Niemela, and R. H., Miller. 1989. One-step preparation of competent *Escherichia coli*: transformation and storage of bacterial cells in the same solution. *Proceedings of the National Academy of Sciences* 86: 2172–2175

Ciuffi, A., and F. D., Bushman. 2006. Retroviral DNA integration: HIV and the role of LEDGF/p75. *Trends in Genetics* 22: 388–395

Clark, L., and J., Carbon. 1976. A colony bank containing synthetic *ColE1* hybrids representative of the entire *E. coli* genome. *Cell* 9: 91–99

Cline, J., J. C., Braman, and H. H., Hogrefe. 1996. PCR fidelity of *Pfu* DNA polymerase and other thermostable DNA polymerases. *Nucleic Acids Research* 24: 3546–3551

Cohen, S. N., A. C., Chang, H. W., Boyer, and R. B., Helling. 1973. Construction of biologically functional bacterial plasmids in vitro. *Proceedings of the National Academy of Sciences* 70(11): 3240

Collins, J., and H. J., Bruning. 1978. Plasmids useable as gene-cloning vectors in an in vitropackaging by coliphage lambda: 'cosmids'. *Gene* 4: 85–107

Cooper, D. N., and J. F., Clayton. 1988. DNA polymorphism and the study of disease associations. *Human Genetics* 78(4): 299–312

Cowling, V. H. 2009. Regulation of mRNA cap methylation. *Biochemical Journal* 425: 295–302

Crick, F. H. C., and J. D., Watson. 1954. The complementary structure of deoxyribonucleic acid. *Proceedings of the Royal Society, London* 223: 80–96

Dagert, M., and S. D., Ehrlich. 1974. Prolonged incubation in calcium chloride improves competence of *Escherichia coli* cells. *Gene* 6: 23–28

Dame, R. T. 2005. The role of nucleoid-associated proteins in the organization and compaction of bacterial chromatin. *Molecular Microbiology* 56: 858–870

Das, A. 1993. Control of transcription termination by RNA-binding proteins. *Annual Review of Biochemistry* 62: 893–930

Davey, M. J., and M., O'Donnell. 2000. Mechanisms of DNA replication. *Current Opinion in Chemical Biology* 4: 581–586

de Boer, H. A., L. J., Comstock, and M., Vasser. 1983. The tacpromoter: a functional hybrid derived from the *trp* and *lac* promoters. *Proceedings of the National Academy of Sciences* 80: 21–25

deHaseth, P. L., M. L., Zupancic, and M. T. Jr., Record. 1998. RNA polymerase-promoter interactions: the comings and goings of RNA polymerase. *Journal of Bacteriology* 180: 3019–3025

Doherty, A. J., S. R., Ashford, H. S., Subramanya, and D. B., Wigley. 1996. Bacteriophage T7 DNA ligase. Overexpression, purification, crystallization, and characterization. *Journal of Biological Chemistry* 271: 11 083–11 089

Dower, W. J., J. F., Miller, and C. W., Ragsdale. 1988. High efficiency transformation of *E. coli* by high voltage electroporation. *Nucleic Acids Research* 16: 6127–6145

Dryden, D. T., N. E., Murray, and D. N., Rao. 2001. Nucleoside triphosphate-dependent restriction enzymes. *Nucleic Acids Research* 29: 3728–3741

Dubendorff, J. W., and F. W., Studier. 1991. Controlling basal expression in an inducible T7 expression system by blocking the target T7 promoter with lac repressor. *Journal of Molecular Biology* 219: 45–59

Eisenstein, B. I. 1990. The polymerase chain reaction: a new method for using molecular genetics for medical diagnosis. *New England Journal of Medicine* 322: 178–183

Endo, Y., and T., Sawasaki. 2006. Cell-free expression systems for eukaryotic protein production. *Current Opinion in Biotechnology* 17: 373–380

Fangman, W. L. 1978. Separation of very large DNA molecules by gel electrophoresis. *Nucleic Acids Research* 5: 653–665

Fire, A., S., Xu, M. K., Montgomery, S. A., Kostas, S. E., Driver, and C. C., Mello. 1998. Potent and specific genetic interference by double-stranded RNA in *Caenorhabditis elegans*. *Nature* 391: 806–811

Freed, E. O. 2001. HIV-1 replication. *Somatic Cell and Molecular Genetics* 26(1–6): 13–33

Frick, D. N., and C. C., Richardson. 2001. DNA primases. *Annual Review of Biochemistry* 70: 39–80

Frischauf, A. M., H., Lehrach, A., Poustka, and N., Murray. 1983. Lambda replacement vectors carrying polylinker sequences. *Journal of Molecular Biology* 170: 827–842

Foley, K. P., M. W., Leonard, and J. D., Engel. 1993. Quantitation of RNA using the polymerase chain reaction. *Trends in Genetics* 9: 380–386

Fuchs, R., and R., Blakesley. 1983. Guide to the use of type II restriction endonucleases. *Methods in Enzymology* 100: 3–38

Fuchs, R. P., S., Fujii, and J., Wagner. 2004. Properties and functions of *Escherichia coli:* Pol IV and Pol V. *Advances in Protein Chemistry* 69: 229–264

Gefter, M. L. 1975. DNA replication. *Annual Review of Biochemistry* 874: 45–74

Ghosh, A., and M., Bansal. 2003. A glossary of DNA structures from A to Z. *Acta Crystallographica Section D: Biological Crystallography* 59: 620–626

Gibson, T. J., A. R., Coulson, J. E., Sulston, and P. F. R., Little. 1987. Lorist 2, a cosmid with transcriptional terminators insulating vector genes from interference by promoters within the insert: effect on DNA yield and cloned insert frequency. *Gene* 53: 275–281

Glover, B. P., and C. S., McHenry. 2001. The DNA polymerase III holoenzyme: an asymmetric dimeric replicative complex with leading and lagging strand polymerases. *Cell* 105: 925–934

Gogarten, J. P., A. G., Senejani, O., Zhaxybayeva, L., Olendzenski, and E., Hilario. 2002. Inteins: structure, function, and evolution. *Annual Review of Microbiology* 56: 263–287

Gomez, C., and T. J., Hope. 2005. The ins and outs of HIV replication. *Cell Microbiology* 7: 621–626

Green, P. J., O., Pines, and M., Inouye. 1986. The role of antisense RNA in gene regulation. *Annual Review of Biochemistry* 55: 569–597

Green, E. D., H. C., Riethman, J. E., Dutchik, and M.V., Olson. 1991. Detection and characterization of chimeric yeast artificial-chromosome clones. *Genomics* 11: 658–669

Griess, G., A., D., Louie, and P., Serwer. 1995. A desktop, low-cost video fluorometer for quantitation of macromolecules after gel electrophoresis. *Applied and Theoretical Electrophoresis* 4: 175–177

Griffin, H. G., and A. M., Griffin. 1993. Dideoxy sequencing reactions using Sequenase version 2.0. *Methods in Molecular Biology* 23: 103–108

Grivitz, S. C., S., Bacchetti, A. J., Rainbow, and F. L., Graham. 1980. A rapid and efficient procedure for the purification of DNA from agarose gels. *Analytical Biochemistry* 106: 492–496

Grunstein, M., and D. S., Hogness. 1975. Colony hybridization: a method for the isolation of cloned DNAs that contain a specific gene. *Proceedings of the National Academy of Sciences* 72: 3961–3965

Grunstein, M., and J., Wallis. 1979. Colony hybridization. *Methods in Enzymology* 68: 379–389

Gubler, U., and B. J., Hoffman. 1983. A simple and very efficient method for generating cDNA libraries. *Gene* 25: 263–269

Hager, G. L., J. G., McNally, and T., Misteli. 2009. Transcription dynamics. *Molecular Cell* 35: 741–753

Hakem, R. 2008. DNA—damage repair; the good, the bad, and the ugly. *EMBO Journal* 27: 589–605

Hanahan, D. 1983. Studies on transformation of *Escherichia coli* with plasmids. *Journal of Molecular Biology* 166: 557–580

Harlow, E., and D., Lane. 1999. *Using Antibodies: a laboratory manual*. New York: Cold Spring Harbour Laboratory Press.

Harrington, J. J., G., Van Bokkelen, R. W., Mays, K., Gustashaw, and H. F., Willard. 1997. Formation of *de novo* centromeres and construction of first-generation human artificial microchromosomes. *Nature Genetics* 15: 345–355

Heidecker, G., J., Messing, and B., Gronenborn. 1980. A versatile primer for DNA sequencing in the M13mp2 cloning system. *Gene* 10(1): 69–73

Helling, R. B., H. M., Goodman, and H. W., Boyer. 1974. Analysis of endonuclease R-EcoRI fragments of DNA from lamboid bacteriophages and other viruses by agarose-gel electrophoresis. *Journal of Virology* 14: 1235–1244

Hendrix, R., J., Roberts, F., Stahl, and R., Weisberg. 1983. *Lambda II*. New York: Cold Spring Harbour Laboratory Press

Henkin, T. M. 1996. Control of transcription termination in prokaryotes. *Annual Review of Genetics* 30: 35–57

Higashitani, N., A., Higashitani, and K., Horiuchi. 1995. SOS induction in *E. coli* by single stranded DNA of mutant filamentous phage: monitoring by cleavage of LexA repressor. *Journal of Bacteriology* 177: 3610–3612

Higashitani, A., N., Higashitani, and K., Horiuchi. 1997. Minus-strand origin of filamentous phage versus transcriptional promoters in recognition of RNA polymerase. *Proceedings of the National Academy of Sciences* 1(94): 2909–2914

Higuchi, R., B., Krummel, and R., Saiki. 1988. A general method of *in vitro* preparation and specific mutagenesis of DNA fragments: study of protein and DNA interactions. *Nucleic Acids Research* 16: 7351–7367

Holmes, D. S., and M., Quigley. 1981. A rapid boiling method for the preparation of bacterial plasmids. *Analytical Biochemistry* 114: 193–197

Hope, I. A. 2001. RNAi surges on: application to cultured mammalian cells. *Trends in Genetics* 17: 440

Horiuchi, K. 1997. Initiation mechanisms in replication of filamentous phage DNA. *Genes to Cells* 2: 425–432

Houdebine, L. M., and J., Attal. 1999. Internal ribosome entry sites (IRESs): reality and use. *Transgenic Research* 8: 157–177

Hsu, L. M. 2002. Promoter clearance and escape in prokaryotes. *Biochimicaet Biophysica Acta* 1577: 191–207

Hutvagner, G., and P. D., Zamore. 2002. RNAi: nature abhors a double strand. *Current Opinion in Genetics and Development* 12: 225–232

Hutvagner, G., J., McLachlan, A. E., Pasquinelli, E., Balint, T., Tuschl, and P. D., Zamore. 2001. A cellular function for the RNA-interference enzyme Dicer in the maturation of the *let*-7 small temporal RNA. *Science* 293: 834–838

Huynh, T. V., R. A., Young, and R. W., Davis. 1985. Construction and screening of cDNA libraries in lgt10 and lgt11. In *DNA Cloning*, Vol. 1: *A practical approach*, pp. 49–78, edited by D. M. Glover. Oxford: IRL Press

Innis, M. A., D. H., Gelfand, and J. J., Sninsky (eds). 1999. *PCR Applications: protocols for Functional Genomics*. San Diego: Academic Press

Ish-Horowitz, D., and J. F., Burke. 1981. Rapid and efficient cosmid vector cloning. *Nucleic Acids Research* 9: 2989–2999

Jeffreys, A. J., V., Wilson, and S. W., Thein. 1984. Hypervariable 'minisatellite' regions in human DNA. *Nature* 314: 67–73

Johnson, I. S. 1983. Human insulin from recombinant DNA technology. *Science* 219(4585): 632–637

Kamath, R. S., A. G., Fraser, Y., Dong, G., Poulin, R., Durbin, M., Gotta, A., Kanapin, N., Le Bot, S., Moreno, M., Sohrmann, D. P., Welchman, P., Zipperlen, and J., Ahringer. 2003. Systematic functional analysis of the *Caenorhabditiselegans* genome using RNAi. *Nature* 421: 231–237

Karn, J., S., Brenner, L., Barnett, and G., Cesareni. 1980. Novel bacteriophage lambda cloning vector. *Proceedings of the National Academy of Sciences* 77: 5172–5176

Kellogg, D. E., I., Rybalkin, S., Chen, N., Mukhamedova, T., Vlasik, P. D., Siebert, and A., Chenchik. 1994. TaqStart Antibody: 'hot start' PCR facilitated by a neutralizing monoclonal antibody directed against Taq DNA polymerase. *Biotechniques* 16: 1134–1137

Keohavong, P., and W. G., Thilly. 1989. Fidelity of DNA polymerases in DNA amplification. *Proceedings of the National Academy of Sciences* 86: 9253–9257

Keohavong, P, L., Ling, C., Dias, and W. G., Thilly. 1993. Predominant mutations induced by the *Thermococcuslitoralis*, vent DNA polymerase during DNA amplification in vitro. *PCR Methods and Applications* 2: 288–292

Kirkpatrick, F. H. 1990. Overview of agarose gel properties. *Current Communications in Cell and Molecular Biology* 1: 9–22

Kneller, E. L., A. M., Rakotondrafara, and W. A., Miller. 2006. Cap-independent translation of plant viral RNAs. *Virus Research* 119: 63–75

Kolitz, S. E., and J. R., Lorsch. 2010. Eukaryotic initiator tRNA: finely tuned and ready for action. *FEBS Letters* 584: 396–404

Kovall, R. A., and B. W., Matthews. 1999. Type II restriction endonucleases: structural, functional, and evolutionary relationships. *Current Opinion in Chemical Biology* 3: 578–583

Kozak, M. 2002. Pushing the limits of the scanning mechanism for initiation of translation. *Gene* 299: 1–34

Kozak, M. 2005. Regulation of translation via mRNA structure in prokaryotes and eukaryotes. *Gene* 361: 13–37

Kunkel, T. A. 1985. Rapid and efficient site-specific mutagenesis without phenotypic selection. *Proceedings of the National Academy of Sciences* 82: 488–492

Kunkel, T. A. 1987. Rapid and efficient site-specific mutagenesis without phenotypic selection. *Methods in Enzymology* 154: 367–383

Lewin, B. 2004. *Genes VIII*. New Jersey: Pearson Education Inc.

Ling, M. L., S. S., Risman, J. F., Klement, N., McGraw, and W. T., McAllister. 1989. Abortive initiation by bacteriophage T3 and T7 polymerases under conditions of limiting substrate. *Nucleic Acids Research* 17: 1605–1618

Lodish, H., A., Berk, S. L., Zipursky, P., Matzudaira, D., Baltimore, and J., Darnell. 1999. *Molecular Cell Biology*, 4th edn. New York: WH Freeman and Co.

Makrides, S. C. 1996. Strategies for achieving high-level expression of genes in *Escherichia coli*. *Microbiological Reviews* 60: 512–538

Marchuk, D., M., Drumm, A., Saulino, and F. S., Collins. 1991. Construction of T-vectors, a rapid and general system for direct cloning of unmodified PCR products. *Nucleic Acids Research* 19: 1154

Margolis, J., and C. W., Wrigley. 1975. Improvement of pore gradient electrophoresis by increasing the degree of cross-linking at high acrylamide concentration. *Journal of Chromatography* 106: 204–209

Mariño-Ramírez, L., M. G., Kann, B. A., Shoemaker, and D., Landsman. 2005. Histone structure and nucleosome stability. *Expert Review of Proteomics* 2: 719–729

Martin, L. J. 2008. DNA damage and repair: relevance to mechanisms of neurodegeneration. *Journal of Neuropathology and Experimental Neurology* 67: 377–387

Martinez, J., A., Patkaniowska, H., Urlaub, R., Lührmann, and T., Tuschl. 2002. Single stranded antisense siRNAs guide target RNA cleavage in RNAi. *Cell* 110: 563

McDonell, M. W., M. N., Simon, and F. W. Stuier. 1977. Analysis of restriction fragments of T7 DNA and determination of molecular weights by electrophoresis in neutral and alkaline gels. *Journal of Molecular Biology* 110: 119–146

McHenry, C. S. 1988. DNA polymerase III holoenzyme of *Escherichia coli*. *Annual Review of Biochemistry* 57: 519–550

McPherson, M. J., and S. G., Møller(eds). 2000. *PCR Basics: from Background to Bench*. Oxford: BIOS

Meselson, M., and R., Yuan. 1968. DNA restriction enzyme from *E. coli*. *Nature* 217: 1110–1114

Messing, J. 1983. New M13 vectors for cloning. *Methods in Enzymology* 101: 20–78

Mignone, F., C., Gissi, S., Liuni, and G., Pesole. 2002. Untranslated regions of mRNAs. *Genome Biology* 3: reviews0004.0001–reviews0004.0010

Miller, J. H. 1978. The *lac*I gene: its role in lac operon control and its uses as a genetic system. In *The Operon*, pp. 31–88, edited by J. Miller. New York: Cold Spring Harbour Laboratory Press

Mocharla, H., R., Mocharla, and M. E., Hodes. 1990. Coupled reverse transcription–polymerase chain reaction (RT–PCR) as a sensitive and rapid method for isozyme genotyping. *Gene* 93: 271–275

Moffatt, B. A., and F. W., Studier. 1986. Use of bacteriophage T-7 RNA polymerase to direct selective high-level expression of cloned genes [abstract]. *Journal of Molecular Biology* 189(1): 113–130

Mullis, K. B. 1990. The unusual origin of the polymerase chain reaction. *Scientific American* 262(4): 56–65

Mullis, K. B., and F. A., Faloona. 1987. Specific synthesis of DNA in vitro via a polymerase catalysed chain reaction. *Methods in Enzymology* 155: 335–350

Nathans, D., and H. O., Smith. 1975. Restriction endonucleases in the analysis and restructuring of DNA molecules. *Annual Review of Biochemistry* 44: 273–93

Narayanan, S. 1991. Applications of restriction fragment length polymorphism. *Annals of Clinical and Laboratory Science* 21(4): 291–296

Nelson, D. L., and M. M., Cox. 2004. *Lehninger Principles of Biochemistry*, 4th edn. New York: WH Freeman and Co.

Ngo, H., C., Tschudi, K., Gull, and E., Ullu. 1998. Double-stranded RNA induces mRNA degradation in *Trypanosomabrucei*. *Proceedings of the National Academy of Sciences* 95: 14687–14692

Nisole, S., and A., Saïb. 2004. Early steps of retrovirus replicative cycle. *Retrovirology* 1: 9

Norrander, J., T., Kempe, and J., Messing. 1983. Construction of improved M13 vectors using oligodeoxynucleotide-directed mutagenesis. *Gene* 26: 101–106

Nudler, E. 1999. Transcription elongation: structural basis and mechanisms. *Journal of Molecular Biology* 288: 1–12

Ohama, T., Y., Inagaki, Y., Bessho, and S., Osawa. 2008. Evolving genetic code. *Proceedings of the Japan Academy—Series B: Physical and Biological Sciences* 84: 58–74

Osawa, S., T. H., Jukes, K., Watanabe, and A., Muto. 1992. Recent evidence for evolution of the genetic code. *Microbiological Reviews* 56: 229–264

Patel, P. H., M., Suzuki, E., Adman, A., Shinkai, and L. A., Loeb. 2001. Prokaryotic DNA polymerase I: evolution, structure, and 'base flipping' mechanism for nucleotide selection. *Journal of Molecular Biology* 308: 823–837

Pingoud, A., and A., Jeltsch. 2001. Structure and function of type II restriction endonucleases. *Nucleic Acids Research* 29: 3705–3727

Ptashne, M. 1992. *A Genetic Switch: phage l and higher organisms.* Cambridge: Cell Press–Blackwell

Radloff, R., W., Bauer, and J., Vinograd. 1967. A dye-buoyant-density method for the detection and isolation of closed circular duplex DNA: The closed circular DNA in HeLa cells. *Proceedings of the National Academy of Sciences* 57: 1514–1521

Raj Bhandary, U. L. 1994. Initiator transfer RNAs. *Journal of Bacteriology* 176: 547–552

Ramakrishnan, V. 2002. Ribosome structure and the mechanism of translation. *Cell* 108: 557–572

Raymond, S., and L., Weintraub. 1959 Acrylamide gel as a supporting medium for zone electrophoresis. *Science* 130: 711

Riemer, J., N., Bulleid, and J. M., Herrmann. 2009. Disulfide formation in the ER and mitochondria: two solutions to a common process. *Science* 5(324): 1284–1287

Roberts, R. J. 1994. Restriction enzymes and their isoschizomers. *Nucleic Acids Research* 20 (Supp.#1): 2167–2180

Roberts, R. J., and D., Macelis. 1997. REBASE—restriction enzymes and methylases. *Nucleic Acids Research* 25: 248–262

Roberts, R. J., M., Belfort, and T., Bestor, *et al.* 2003. A nomenclature for restriction enzymes, DNA methyltransferases, homing endonucleases and their genes. *Nucleic Acids Research* 31: 1805–1812

Rothwell, P. J., and G., Waksman G. 2005. Structure and mechanism of DNA polymerases. *Advances in Protein Chemistry* 71: 401–440

Rudner, R., B., Studamire, and E. D., Jarvis. 1994. Determinations of restriction fragment length polymorphism in bacteria using ribosomal RNA genes. *Methods in Enzymology* 235: 184–196

Saiki, R. K., S., Scharf, F., Faloona, K. B., Mullis, H. A., Erlich, and N., Arnheim. 1985. Enzymatic amplification of beta-globin genomic sequences and restriction site analysis for diagnosis of sickle cell anemia. *Science* 230(4732): 1350–1354

Saltzman, A. G., and R., Weinmann. 1989. Promoter specificity and modulation of RNA polymerase II transcription. *FASEB Journal* 3: 1723–1733

Sandman, K., and J. N., Reeve. 2000. Structure and functional relationships of archaeal and eukaryal histones and nucleosomes. *Archives of Microbiology* 173: 165–169

Sanger, F., A. R., Coulson, B. G., Barrell, A. J. M., Smith, and B. A., Roe. 1980. Cloning in single stranded bacteriophage as an aid to rapid DNA sequencing. *Journal of Molecular Biology* 143: 161–178

Sanger, F., S., Nicklen, and A. R., Coulson. 1977. DNA sequencing with chain-terminating inhibitors. *Proceedings of the National Academy of Sciences* 74: 5463–5467

Schärer, O. D. 2003. Chemistry and biology of DNA repair. *Angewandte Chemie (International ed. in English)* 42: 2946–2974

Schmeing, T. M., and V., Ramakrishnan. 2009. What recent ribosome structures have revealed about the mechanism of translation. *Nature* 461: 1234–1242

Schwartz, D. C., and C. R., Cantor. 1984. Separation of yeast chromosome-sized DNA by pulsed field gel electrophoresis. *Cell* 37: 67–75

Seed, B., R. C., Parker, and N., Davidson. 1982. Representation of DNA sequences in recombinant DNA libraries prepared by restriction enzyme partial digestion. *Gene* 19: 201–209

Shao, Y., and S. B., Kent. 1997. Protein splicing: occurrence, mechanisms, and related phenomena. *Chemistry and Biology* 4: 187–194

Sharp, P. A., B., Sugden, and J., Sambrook. 1973. Detection of two restriction endonuclease activities in *Haemophilus parainfluenzae* using analytical agarose-ethidium bromide electrophoresis. *Biochemistry* 12: 3055–3063

Shimizu, Y., Y., Kuruma, B. W., Ying, S., Umekage, and T., Ueda. 2006. Cell-free translation systems for protein engineering. *FEBS Journal* 273: 4133–4140

Shizuya, H., B., Birren, U. J., Kim, V., Mancino, T., Slepak, Y., Tachiiri, and M., Simon. 1992. Cloning and stable maintenance of 300-kilobase-pair fragments of human DNA in *Escherichia coli* using an F-factor-based vector. *Proceedings of the National Academy of Sciences, USA* 89: 8794–8797

Short, J. M., J. M., Fernandez, J. A., Sorge, and W. D., Huse. 1988. lZAP: a bacteriophage lambda expression vector with in vivoexcision properties. *Nucleic Acids Research* 16: 7583–7600

Sims, R. J., R., Belotserkovskaya, and D., Reinberg. 2004. Elongation by RNA polymerase II: the short and long of it. *Genes and Development* 18: 2437–2468

Slater, G. W., P., Mayer, and G., Drouin. 1996. Migration of DNA through gels. *Methods in Enzymology* 270: 272–295

Smith, L., J., Sanders, R., Kaiser, P., Hughes, C., Dodd, C., Heiner, S., Kent, and L., Hood. 1986. Fluorescence detection in automated DNA sequence analysis. *Nature* 321: 674–679

Stillman, B. 1989. Initiation of eukaryotic DNA replication in vitro. *Annual Review of Cell Biology* 5: 197–245

Temin, H. M., and S., Mizutani. 1970. RNA-dependent DNA polymerase in virions of Rous sarcoma virus. *Nature* 226: 1211–1213

Tenson, T., and V., Hauryliuk. 2009. Does the ribosome have initiation and elongation modes of translation? *Molecular Microbiology* 72: 1310–1315

Terry, B (ed.). 1991. *DNA Fingerprinting: approaches and applications*. Boston: Birkhauser Verlag. [Papers from the First International Symposium on DNA Fingerprinting, Bern, Switzerland, October 1990]

Timson, D. J., M. R., Singleton, and D. B., Wigley. 2000. DNA ligases in the repair and replication of DNA. *Mutation Research* 460: 301–318

Tost, J. 2010. DNA methylation: an introduction to the biology and the disease-associated changes of a promising biomarker. *Molecular Biotechnology* 44: 71–81

Tremphy, J. E., and T., Nancy. 2003. Gene expression and regulation. In *Fundamental Bacterial Genetics*. UK: Blackwell Publishing

Vieira, J., and J., Messing. 1982. The pUC plasmids, an M13mp7-derived system for insertion mutagenesis and sequencing with synthetic universal primers. *Gene* 19: 259–268

Vologodskii, A. V., and N. R., Cozzarelli. 1994. Conformational and thermodynamic properties of supercoiled DNA. *Annual Review of Biophysics and Biomolecular Structure* 23: 609–643

Wachtel, C., and J. L., Manley. 2009. Splicing of mRNA precursors: the role of RNAs and proteins in catalysis. *Molecular BioSystems* 5: 311–316

Waga, S., and B., Stillman. 1998. The DNA replication fork in eukaryotic cells. *Annual Review of Biochemistry* 67: 721–751

Wambaugh, J. 1989. *The Blooding*. NY: Morrow (novel about the first use of DNA fingerprinting in a court case)

Waring, M. J. 1965. Complex formation between ethidium bromide and nucleic acids. *Journal of Molecular Biology* 13: 269–282

Webb, R., K. J., Reddy, and L. A., Sherman. 1989. Lambda ZAP: Improved strategies for expression library construction and use. *DNA* 8: 69–73

Weiss, B., and C. C., Richardson. 1967. Enzymatic breakage and joining of deoxyribonucleic acid, I. Repair of single-strand breaks in DNA by an enzyme system from *Escherichia coli* infected with T4 bacteriophage. *Proceedings of the National Academy of Sciences* 57: 1021–1028

Weiss, R. A. 1996. Retrovirus classification and cell interactions. *Journal of Antimicrobial Chemotherapy* 37: 1–11

Wells, J. A., M., Vasser, and D. B., Powers. 1985. Cassette mutagenesis: an efficient method for generation of multiple mutations at defined sites. *Gene* 34: 315–323

Wittwer, C. T., M. G., Herrmann, Moss, A. A., and R. P., Rasmussen. 1997. Continuous fluorescence monitoring of rapid cycle DNA amplification. *Biotechniques* 22: 130–138

Yanisch-Perron, C., J., Vieira, and J., Messing. 1985. Improved M13 phage cloning vectors and host strains: nucleotide sequences of the M13mp18 and pUC19 vectors. *Gene* 33: 103–119

Yansura, D. G., and D. J., Henner. 1990. Use of *Escherichia coli* trp promoter for direct expression of proteins. *Methods in Enzymology* 185: 54–60

Young, R. A., and R. W., Davis. 1983. Efficient isolation of genes by using antibody probes. *Proceedings of the National Academy of Sciences* 80: 1194–1198

Zhao, J., L., Hyman, and C., Moore. 1999. Formation of mRNA 3′-ends in eukaryotes: mechanism, regulation, and interrelationships with other steps in mRNA synthesis. *Microbiology and Molecular Biology Reviews* 63: 405–445

Zhou, M. Y., S. E., Clark, and C. E., Gomez-Sanchez. 1995. Universal cloning method by TA strategy. *Biotechniques* 19: 34–35

Zimmerman, S. B., J. W., Little, C. K., Oshinsky, M., Gellert. 1967. Enzymatic joining of DNA strands: a novel reaction of diphosphopyridine nucleotide. *Proceedings of the National Academy of Sciences* 57: 1841–1848

Zoller, M. J., and M., Smith. 1983. Oligonucleotide-directed mutagenesis of DNA fragments cloned into M13 vectors. *Methods in Enzymology* 100: 468–500

Websites

<www.ambion.com/techlib/basics/translation/index.html>

<www-genome.wi.mit.edu/genome software/other/primer3.html>

Glossary

Active site   A part of an enzyme where substrates bind and undergo a chemical reaction. It is usually found in a cleft or pocket covered by an amino acid residue, which recognizes the appropriate substrate. The substrate has the ability to bind to these sites by different interactions like hydrogen and van der Waals bonds, which result in the formation of an enzyme–substrate complex.

Amino acids   Structural units of proteins known as building blocks. They consist of an amino group, a carboxyl group, a hydrogen atom, and an R group bonded to the alpha carbon. They also play an indirect role in metabolism. The 20 amino acids found within proteins provide chemical versatility to different proteins. They are classified into acidic, basic, hydrophobic, and polar amino acids.

Amplicon   Specific DNA product generated by polymerase chain reaction (PCR) using a pair of PCR primers.

Antibodies   A protein produced by the body's immune system (by a type of white blood cells called B-cells) against foreign particles (antigens) considered harmful to the body and are sensitized by the immune system. These antigens can be exemplified as various microorganisms like bacteria, fungi, parasites, viruses, and some chemicals. Each antibody is unique and is released against specific antigens as programmed by the immune system. These specialized immunoglobulins are released on the introduction of an antigen in the body and attach with that antigen.

Bacteria   Microscopic prokaryotes devoid of membrane-enclosed nucleus and other organelles like mitochondria and chloroplasts. They have rigid cell walls made up of peptidoglycans. They also have a closed circular double-stranded DNA with no associated histones. Sometimes, flagella are also present.

Bacteriophage   Obligate intracellular parasites living inside bacterial cells and multiplying within. They exploit the host by overtaking the host's biosynthetic machinery.

cDNAs   Single-stranded DNA synthesized from mRNA templates with the help of enzyme reverse transcriptase by a process known as reverse transcription (RT) or first-strand cDNA synthesis. They are used for analysing mRNAs and are important because they are comparatively more stable.

Cells   Structural and functional units of living organisms, consisting of all cellular organelles. They take in nutrients and convert them into energy, which is used in carrying out functions like metabolism and reproduction.

Chromatin   Mass genetic material composed of DNA and proteins; present in the nucleus and condenses furthermore to form chromosomes in the course of cell division.

Chromosomes   Present in the nucleus of the cell; basically thread-like structures packaged with DNA. Each chromosome is made up of tightly coiled DNA (containing genes and regulatory elements) around histone proteins, which gives stability to its structure and also controls its functions.

Codon   A sequence of three adjacent nucleotides that constitute into a genetic code. It specifies individual and specific amino acids to be added in the polypeptide chain during protein synthesis. The order of the codons along the DNA or messenger RNA is used for deducing the amino acid sequence in the polypeptide chain.

Cytoplasm   The cell content surrounding the nucleus and is contained within the plasma membrane. It is a jelly like lattice, which interconnects and supports solid structures. It consists of cytosol (soluble portion) and cytoplasmic filaments in case of eukaryotes.

Disulphide bond   A type of strong covalent bond (−S−S−), which is important in linking polypeptide chains in proteins. This linkage is formed by the oxidation of sulfhydryl (SH) groups of two molecules of cysteine, resulting into a strong disulphide bond.

DNA polymerase   A heat-resistant enzyme that synthesizes new strands of DNA complementary to the template sequence. The most common examples are *Taq* DNA polymerase (derived from *Thermus aquaticus*), and *Pfu* DNA polymerase (derived from *Pyrococcus furiosus*).

DNA probe   Probes are laboratory synthesized with a sequence complementary to the target DNA sequence and are labelled segments of DNA or RNA, which find their importance in searching for the specific sequence of nucleotides in a DNA molecule.

DNA template   A sample DNA containing the target sequence; serves as a base according to which a complementary strand is synthesized by the polymerases.

DNA   Hereditary material in humans and almost all other organisms. DNA is located in the cell nucleus, but a small amount is also found in the mitochondria, called mitochondrial DNA. It stores genetic information in code language consisting of four chemical bases: adenine (A), guanine (G), cytosine (C), and thymine (T).

Enzymes   Proteinaceous biological catalyst molecules having the basic function of increasing the rate of biological reaction and complex with substrates to produce products.

Eukaryotes   Refer to the higher taxonomic kingdom that includes organisms composed of one or more cells containing visibly evident nuclei and organelles.

Gene expression   Process by which information is derived from the gene and is further used in the synthesis of gene products, which are generally proteins (functional RNAs in case of tRNA or rRNA).

Gene silencing   Interruption or suppression of the expression of a gene at transcriptional or translational levels.

Gene   A unit of heredity coded by nucleotides on a stretch of DNA, which codes for a type of protein. Genes have information to build and maintain cells and pass genetic traits to offspring.

Genome   Contains all biological information necessary for growth. The biological information is encoded in DNA and is divided into genes.

HIV (Human Immunodeficiency Virus)   A retrovirus causing AIDS (Acquired Immuno Deficiency Syndrome) by infecting the helper T-cells of the immune system. The most common serotype, HIV-1, is distributed worldwide.

Hormone   A chemical messenger released by one or more cells whose region of action is distant from the region of release, which affects cells in other parts of the organism and is required in very minimal amount. It transports a signal from one cell to another.

Hybridization   Process of annealing of two complementary nucleic acid strands resulting in the formation a double-stranded molecule. It is an important technique for the detection of specific nucleic acid sequences.

Hydrogen bond   Formed due to attractive intermolecular forces that exist between two partial electric charges of opposite polarities.

Immune response   Any reaction by the immune system against the foreign particle.

Immune system   A complex system that distinguishes itself from foreign substances and acts accordingly to protect the human body against foreign invasion by secretly targeting antibodies.

Inhibitors   Molecules interacting with the enzyme, disturbing its normal functioning.

Junk DNA   Non-coding DNA sequences that do not encode for protein sequences. Recent studies have shown them to have biological functions such as transcriptional and translational regulation of protein-coding sequences.

Matrix   A ground substance or medium usually meshwork in which lumps of coarser material such as an aggregate can be embedded.

MicroRNA   Post-transcriptional regulators (~22 nucleotide RNA sequences) binding to complementary sequences in the 3′UTR of multiple target mRNAs and resulting in their silencing. MicroRNAs are able to repress hundreds of targets.

Molecular chaperones   Proteins adapted to have important involvement in protein folding and stabilizing proteins under stress conditions and help in recovering from damage.

Molecular cloning   The transfer of DNA fragment of interest from one cell to a self-replicating plasmid, which is then propagated in a foreign host cell.

Molecule   The smallest unit of a substance that exists single and still retains the character of that particular substance.

Nuclear pore   A protein-covered channel across the nuclear envelope that facilitates the transportation of molecules between the nucleus and the cytoplasm.

Nucleocapsid   Nucleic acid and the surrounding protein coat are together termed nucleocapsid in virus.

Nucleosides   Basic building blocks of nucleic acids formed by the loss of water from sugar and base.

Nucleotides   Subunit of DNA or RNA; several of them together form a DNA or an RNA molecule. Each nucleotide is made up of a nitrogenous base, a phosphate molecule, and a sugar molecule.

Nucleus   A membrane-bound organelle consisting of nucleic acids within, which communicates with the surrounding cytosol via numerous nuclear pores.

Polymerase Chain Reaction (PCR)   Developed by Kary Mullis in the 1980s; a technique for synthesizing new strand of DNA complementary to template by utilizing DNA polymerase. Use of primers makes it possible to amplify only specific region of interest, and hence several copies of amplicons are made at the end of the reaction.

Polysaccharides   Composed of several monosaccharides joined together through dehydration synthesis.

Primers   Small sequences of ssDNA complementary to the target sequence, which act as a focus from where the DNA polymerase binds and starts synthesis of a new strand.

Progeny   A genetic descendant or offspring.

Prokaryotes   Single-celled organisms, which are the most primitive forms of life. Generally, they are non-photosynthetic, but exceptionally, they can have photosynthetic pigments such as cyanobacteria.

Promoter   A segment of DNA present upstream of a gene-coding region, which controls the expression of that gene. A promoter can be used to turn a gene on or off.

Protease   An enzyme that hydrolyses peptide bonds of proteins, and hence called proteolytic enzyme.

Proteins   Large biochemical molecules composed of one or more chains of amino acids. They are required for the structure, function, and regulation of the body's cells, tissues, and organs.

Receptor   A structure on the surface of a cell that selectively binds a specific substrate.

Replication   A process of making an identical copy of a section of duplex (double-stranded) DNA, using existing DNA as a template for the synthesis of new DNA strands. In humans and other eukaryotes, replication occurs in the cell nucleus. When a cell divides, it must first duplicate its genome so that each daughter cell winds up with a complete set of chromosomes.

Reverse transcriptase   A common name for an enzyme that functions as an RNA-dependent DNA polymerase. It is encoded by retroviruses, where it copies the viral RNA genome into DNA prior to its integration into host cells.

Reverse transcription   A process of copying RNA into DNA by the help of an enzyme called reverse transcriptase, and this DNA can be readily integrated into the host cell's genome.

Ribosomes   Synthesize proteins and made up of a large and a small subunit, comprised mostly of ribosomal RNA (rRNA), with proteins interspersed like islands in a sea of RNA. Besides the rRNA, the ribosome contains binding sites for tRNA and mRNA also.

Ribozymes   RNA molecules that catalyse a chemical reaction by the hydrolysis of either one of their own phosphodiester bonds or bonds in other RNAs. They catalyse the aminotransferase activity of the ribosome too.

RNA interference (RNAi)   A conserved biological response to dsRNA that mediates resistance to both the endogenous parasitic and the exogenous pathogenic nucleic acids. It also regulates the expression of protein-coding genes. This natural mechanism may play a vital role in practical applications in functional genomics and therapeutic intervention.

RNA   A molecule consisting of a long chain of nucleotide units; each nucleotide consists of a nitrogenous base, a ribose sugar, and a phosphate group.

Reverse Transcription-PCR (RT-PCR)   A PCR technique converting sample RNA into cDNA with the help of an enzyme called reverse transcriptase.

Sedimentation constant   A measure of the rate at which a molecule suspended in colloidal solution sediments down in an ultracentrifuge. This constant is expressed in Svedberg's units.

Sequencing   A process through which the order of the nucleotides of a gene is determined.

siRNAs   Small interfering RNAs approximately 21–25 base-pair long. Generally, siRNA duplexes are composed of 21 nt sense and 21 nt antisense strands, which pair in a way that they create 2 nt of 3′ overhang, which serves for specific target recognition.

Transcription   Synthesis of RNA using DNA template catalysed by RNA polymerase. The resultant RNA is complementary to the DNA strand.

Transgenic   Transgenic organisms have foreign DNA genes incorporated into their genome during early stages of their development. The transgene is present in both somatic and germ cells and is inherited by offspring.

Transposons   Segments of DNA capable of migrating from the genome of one cell to another, also called "jumping genes", and cause mutations and alter the DNA amount in the genome of cell.

tRNA   Involves in the translation of the nucleic acid into the amino acids of proteins. It has a conserved inverted L structure. One end of the tRNA contains an anticodon loop, which pairs with an mRNA, and the other end of the tRNA attaches to the 3′-OH group by ester linkage.

Vectors   Agents carrying genes from one organism to another; for example, mosquitoes or ticks are vectors that carry disease-causing microorganisms.

Viruses   Submicroscopic parasites of plants, animals, and bacteria that cause diseases and consist of an RNA or a DNA core surrounded by a protein coat. They cannot replicate without host cells.

Colour Plates

Chapter 1
Recombinant DNA Technology

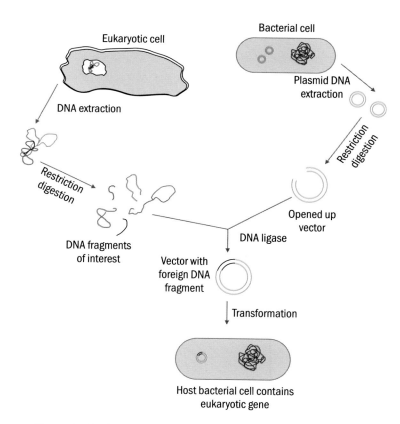

Plate 1 Basic procedure of recombinant DNA technology

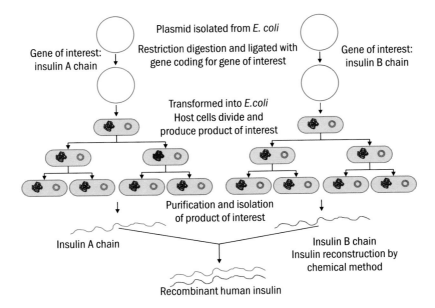

Plate 2   Synthesis of recombinant human insulin in *E. coli*

Chapter 2
Methods for Creating Recombinant DNA Molecules

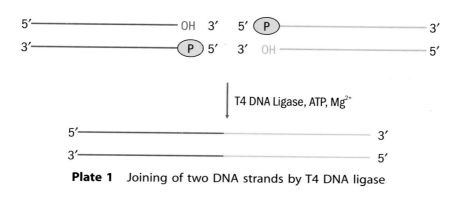

Plate 1   Joining of two DNA strands by T4 DNA ligase

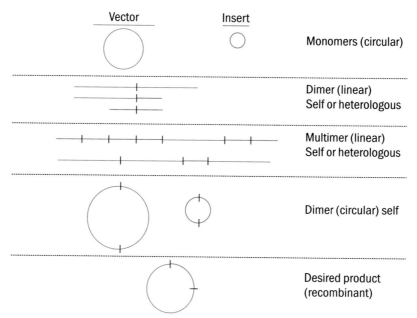

Plate 2   Possible products in ligation reaction

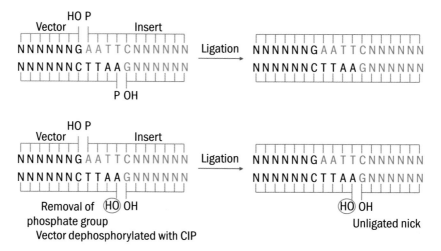

Plate 3   Dephosphorylation of vector and its effect on ligation

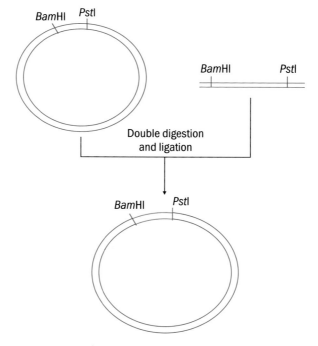

Plate 4   Double digestion

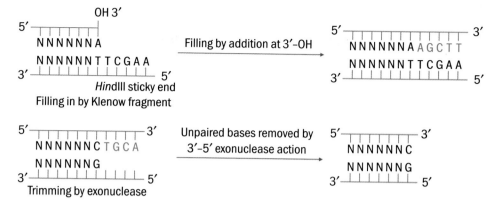

Plate 5   Filling in and trimming

Chapter 5
Construction of DNA Library

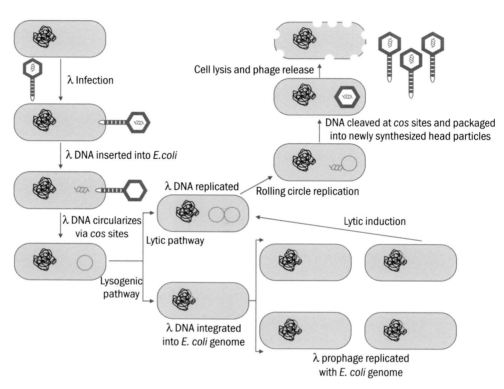

Plate1 λ life cycle

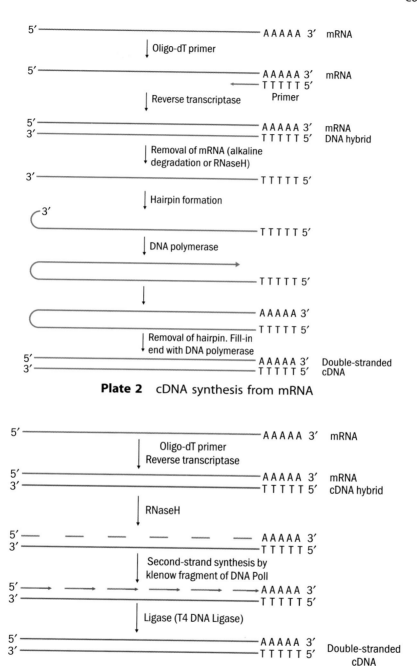

Plate 2 cDNA synthesis from mRNA

Plate 3 cDNA synthesis using RNaseH

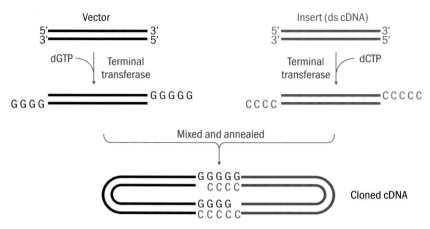

Plate 4   Cloning of cDNA using homopolymer tailing

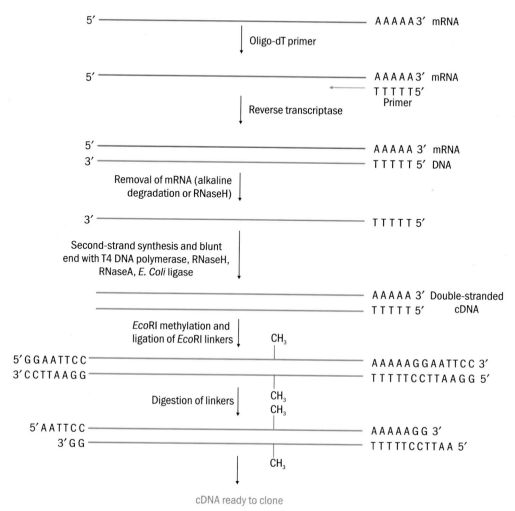

5′ ———————————————————————— A A A A A 3′ mRNA

  Oligo-dT primer

5′ ———————————————————————— A A A A A 3′ mRNA
                                          T T T T T 5′
                                          Primer

  Reverse transcriptase

5′ ———————————————————————— A A A A A 3′ mRNA
3′ ———————————————————————— T T T T T 5′ DNA

  Removal of mRNA (alkaline
  degradation or RNaseH)

3′ ———————————————————————— T T T T T 5′

  Second-strand synthesis and blunt
  end with T4 DNA polymerase, RNaseH,
  RNaseA, *E. Coli* ligase

———————————————————————— A A A A A 3′ Double-stranded
———————————————————————— T T T T T 5′   cDNA

  *Eco*RI methylation and
  ligation of *Eco*RI linkers   CH₃

5′ G G A A T T C C ————————————————— A A A A A G G A A T T C C 3′
3′ C C T T A A G G ————————————————— T T T T T C C T T A A G G 5′

  Digestion of linkers   CH₃
                            CH₃

5′ A A T T C C ————————————————— A A A A A G G 3′
3′ G G ————————————————— T T T T T C C T T A A 5′

              CH₃

cDNA ready to clone

Plate 5 cDNA synthesis from mRNA and preparation for insertion into a vector

Chapter 6
Sequencing by Sanger's Method

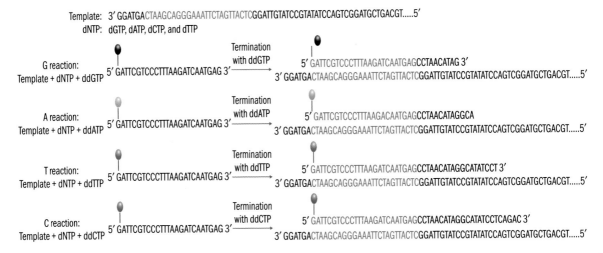

Plate 1 Sequencing reaction using oligonucleotide primers with four different fluorescent labels

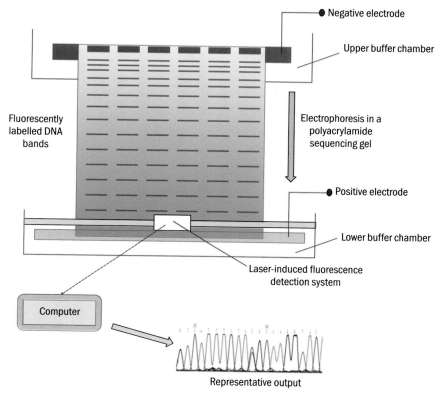

Plate 2   Strategy for automated DNA sequencing

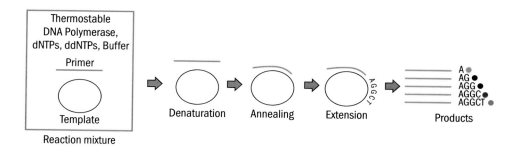

Plate 3   Cycle sequencing

Chapter 7
Protein Production in Bacteria

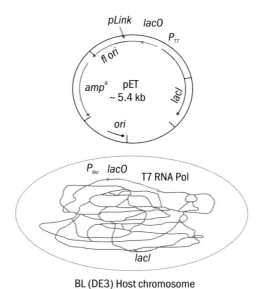

Plate 1 T7-based pET expression system

Chapter 8
Site-directed Mutagenesis

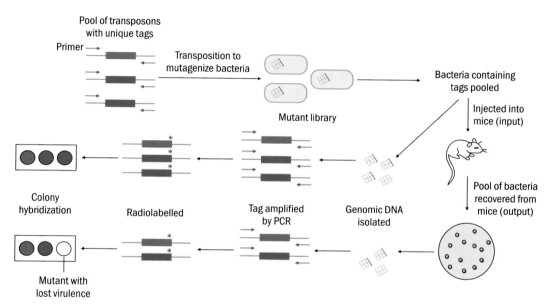

Plate 1 Signature-tagged mutagenesis

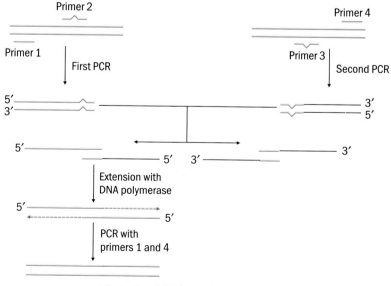

Plate 2 PCR-based mutagenesis

Chapter 10
Polymerase Chain Reaction

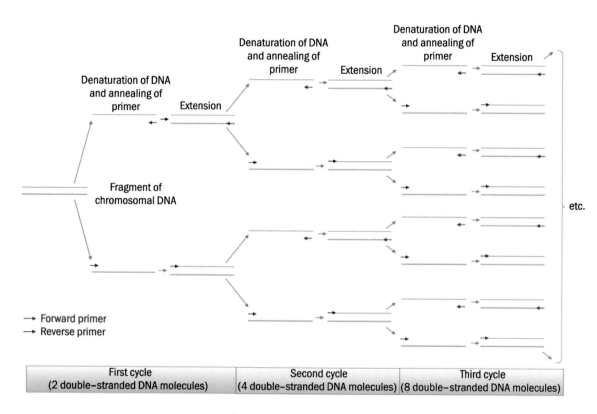

Plate 1 Typical PCR cycles

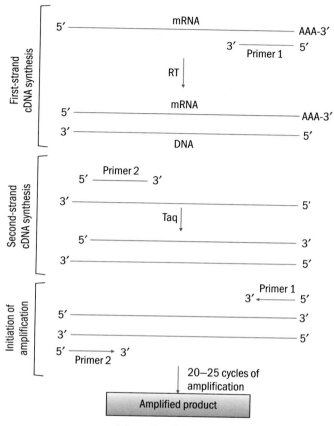

Plate 2   RT–PCR

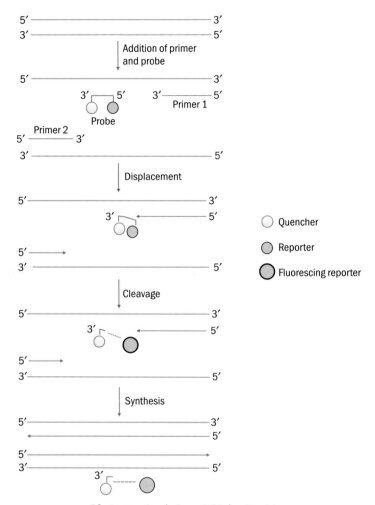

Plate 3 Real-time PCR by TaqMan

Chapter 11
DNA Fingerprinting

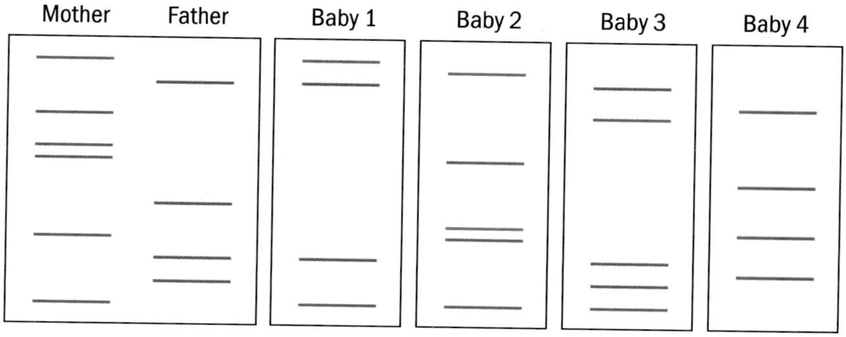

Plate 1   Family testing

Chapter 12
RNAi and siRNA Technology

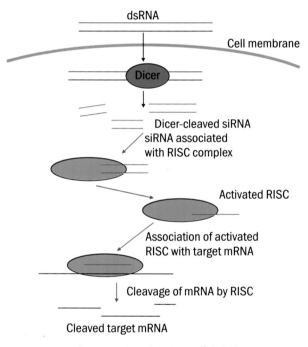

Plate 1 Mechanism of RNAi

Chapter 13
Molecular Biology Methods

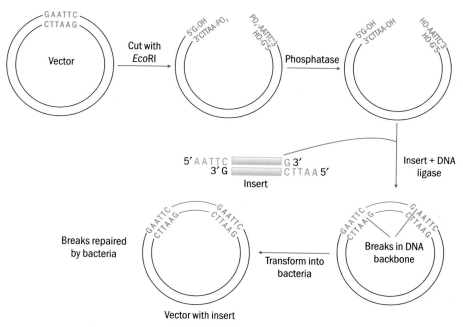

Plate 1 Cloning of fragments with cohesive ends

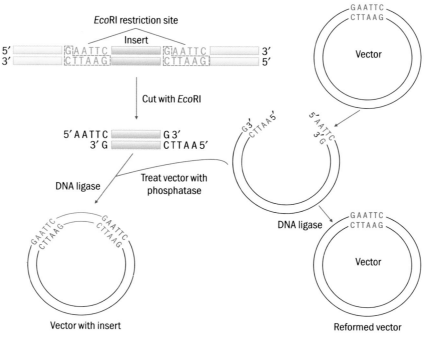

Plate 2   Strategy to protect self-ligation

Chapter 18
Enzymes Used in Molecular Cloning

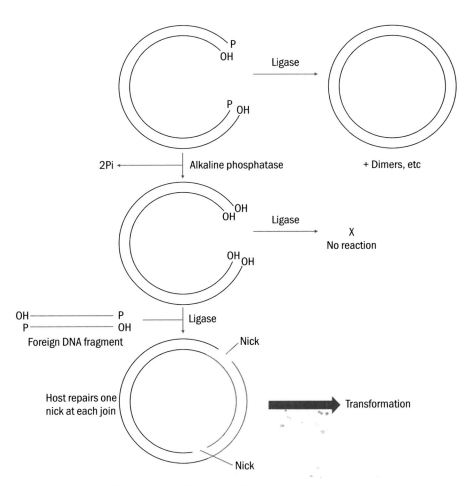

Plate 1 Application of alkaline phosphatase

Chapter 19
Agarose Gel and Polyacrylamide Gel Electrophoresis

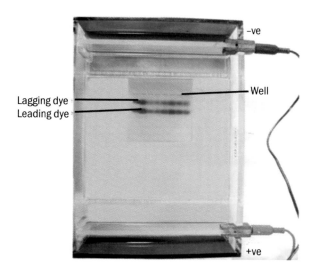

Plate 1   Agarose gel electrophoresis

Index

About the Author

Dr Keya Chaudhuri obtained her PhD from the Jadavpur University in 1983 and since then she has been pursuing a research career as a Scientist at the CSIR–Indian Institute of Chemical Biology (IICB), Kolkata. Presently, she is working as a Chief Scientist in the Molecular and Human Genetics Division of IICB. She has been actively engaged in research in many key areas such as cellular and molecular biology of *Vibrio cholerae*, drug-DNA interactions, gene expression and radiation response, biology of oral cancer and pre-cancer, including the understanding of molecular mechanisms for diagnosis and prevention of disease. She has to her credit about 110 publications in peer-reviewed national and international journals and has co-authored two reference books published by Springer, Berlin.